An
American
Gospel

ALSO BY ERIK REECE

Lost Mountain

AN AMERICAN GOSPEL

On Family,
History,
and the
Kingdom of God

ERIK REECE

RIVERHEAD BOOKS

a member of Penguin Group (USA) Inc. • New York • 2009

RIVERHEAD BOOKS
Published by the Penguin Group
Penguin Group (USA) Inc., 375 Hudson Street, New York, New York 10014, USA •
Penguin Group (Canada), 90 Eglinton Avenue East, Suite 700, Toronto, Ontario
M4P 2Y3, Canada (a division of Pearson Canada Inc.) • Penguin Books Ltd, 80 Strand, London
WC2R 0RL, England • Penguin Ireland, 25 St Stephen's Green, Dublin 2, Ireland
(a division of Penguin Books Ltd) • Penguin Group (Australia), 250 Camberwell Road,
Camberwell, Victoria 3124, Australia (a division of Pearson Australia Group Pty Ltd) •
Penguin Books India Pvt Ltd, 11 Community Centre, Panchsheel Park, New Delhi–110 017, India •
Penguin Group (NZ), 67 Apollo Drive, Rosedale, North Shore 0632, New Zealand (a division
of Pearson New Zealand Ltd) • Penguin Books (South Africa) (Pty) Ltd, 24 Sturdee Avenue,
Rosebank, Johannesburg 2196, South Africa

Penguin Books Ltd, Registered Offices: 80 Strand, London WC2R 0RL, England

The author gratefully acknowledges permission from Wendell Berry
to quote from "Wild Geese" and "Stay Home."

The Moffatt translations are from James Moffatt, *A New Translation of the Bible,
Containing the Old and New Testaments*, 1926; reprinted by Kregel, 1995.

Library of Congress Cataloging-in-Publication Data

Reece, Erik.
An American gospel : on family, history, and the kingdom of God / Erik Reece.
p.   cm.
ISBN 978-1-59448-859-7
1. United States—Religion—History.   2. Christianity—United States.   I. Title.
BL2525.R43   2009   2008054262
200.973—dc22

Printed in the United States of America
1   3   5   7   9   10   8   6   4   2

Book design by Jessica Shatan Heslin/Studio Shatan, Inc.

For my mother

Contents

Introduction   1

I.

First Shot   11

Open Eyes   13

Accommodation   27

William Byrd's New World   37

The Gospel According to Thomas Jefferson   57

II.

Walt Whitman at Furnace Mountain   81

III.

The Kingdom of God   141

The End of Religion   179

Sources   219

Geese appear high over us,
pass, and the sky closes. Abandon,
as in love or sleep, holds
them to their way, clear,
in the ancient faith: what we need
is here. And we pray, not
for new earth or heaven, but to be
quiet in heart, and in eye
clear. What we need is here.

—WENDELL BERRY

An
American
Gospel

Introduction

In the summer of 1890, the world's most famous author, Leo Tolstoy, was reading an essay by a writer the world hardly knew. "The rich man," wrote Henry David Thoreau, "is always sold to the institutions which made him rich." Surrounded by governesses and chambermaids, Tolstoy felt the full sting of that rebuke and found himself wishing he could be as free and unbeholden as the American loner. That same summer, Tolstoy also read "a little book by a very original and courageous poet named Walt Whitman." That little book would have been the first edition of *Leaves of Grass*. Inspired by those two free spirits, Tolstoy began writing a long essay about a third—the Mediterranean wanderer known as Yeshua, whom we today call Jesus. In *The Kingdom of God Is*

Within You, Tolstoy argued that one cannot believe in both the Sermon on the Mount *and* the Nicene Creed.

The Sermon on the Mount begins:

"Blessed are the poor in spirit, for theirs is the kingdom of heaven.

"Blessed are those who mourn, for they shall be comforted.

"Blessed are the meek, for they shall inherit the earth.

"Blessed are those who hunger and thirst for righteousness, for they shall be satisfied.

"Blessed are the merciful, for they shall obtain mercy.

"Blessed are the pure in heart, for they shall see God.

"Blessed are the peacemakers, for they shall be called sons of God.

"Blessed are those who are persecuted for righteousness' sake, for theirs is the kingdom of heaven."

(Matthew 5:2–10, RSV)

In this sermon (repeated as the Sermon on the Plain in the Gospel of Luke), Jesus goes on to charge his crowd to love their enemies, turn the other cheek, give to those who beg, and avoid hypocritical judgments.

By contrast, the Nicene Creed is solely an assertion of the divinity of Jesus, "the only son of God," who was born of a virgin, was crucified as a blood sacrifice for the sins of humankind, and rose from the dead three days later.

Tolstoy argued that either one accepts the Sermon's rigorous demands for how we must act in this world, how we must treat others, or one chooses the Creed as a way of escaping from this world into another. "The man who believes in a god, in a Christ coming again in glory to judge and to punish the quick and the dead," wrote Tolstoy, "cannot believe in the Christ who bade us turn the left cheek, judge not, forgive those that wrong us, and love our enemies." Furthermore, "The man who believes in the Church's doctrine of the compatibility of warfare and capital punishment with Christianity cannot believe in the brotherhood of all men." And finally, "The man who believes in salvation through faith in the redemption or the sacraments cannot devote all his powers to realizing Christ's moral teaching in his life."

The distinction Tolstoy was making came home to me very clearly one morning while I was working on this book. There was a knock at the door. When I answered it, two little girls wearing white dresses were standing on my porch. One asked if they could come in. I looked up and down the street. I didn't see any adults, which made me a little nervous. Who was sending young children door-to-door with no supervision? I told the girls I was actually busy at the moment.

"If you died today," one of them asked, a bit nervously, "are you sure you would go to heaven?"

In my head, I quickly ran through a list of possible answers to this question. None of them seemed very age-appropriate for this audience of two. So finally I said, "Well, I don't think it's quite as simple as that."

The girls looked at each other. An awkward silence ensued. Finally one of them said, "Can we at least leave you something to read?"

I took their leaflet and they walked away. When I opened it, I recognized the photo of a bombastic local preacher. His leaflet read: "There are basically four things you need to know from the Bible." They are:

1. Everyone is a sinner.
2. Everyone owes a sin-debt.
3. Everyone's sin-debt has been paid by the crucifixion of Jesus Christ.
4. Everyone can be saved!

Basically this was all one needed to know. This version of Christianity has nothing to do with following the teachings of the Sermon on the Mount, or with recognizing the kingdom of God within our midst, as Tolstoy advised. It was simply a ticket to heaven. Nothing surprised me about this. I was the son and grandson of Baptist preachers, and this was basically the Christianity I grew up with. But as I would come to believe much later, it isn't the only Christianity, nor is it even the most authentic one.

Exactly fifty years ago, a group of American New Testament scholars presented to the world the Gospel of Thomas, a collection of 114 sayings of Jesus that predates the four canonical Gospels and greatly amplifies Luke 17:21, the verse from which Tolstoy took the title of his book. In the Gospel of Thomas, we hear Jesus' followers ask, "When will the kingdom come?" Jesus responds, "It will not come by watching for it. It will not be said, 'Look, here it is,' or 'Look, there it is.' Rather the father's kingdom is spread out upon the earth, and people do not see it."

Why don't we see it? The natural world is undeniably beautiful. Cherry blossoms are blooming outside my window right now, and tulips are engaged in their perennial resurrection. But we Americans seem to have become too distracted by accumulation and haste to pause over this remarkable observation from the Gospels of Luke and Thomas.

Part of my contention throughout this book is that we Americans are *especially* immune to finding the kingdom of God spread out before us. In the only two states where I have ever lived, Kentucky and Virginia, coal operators, with the full cooperation of the federal government, are blasting the tops off of the oldest mountains in the world, the Appalachians. To get to thin seams of coal, strip miners have leveled 1.5 million acres of mountaintops, all in the name of cheap energy. And this is merely a local example of our country's disastrous refusal to understand the Creation as the work of a divine Creator in whom ninety-five percent of Americans say they believe.

It didn't have to be this way. Christianity did not have to form such an easy and ultimately unholy alliance with industrialism, consumerism, and corporatism. There is another, subversive spirit that runs throughout our history, a strain of thought that provides a religious, ecological, and radically democratic alternative to where we are right now. It is, I believe, a uniquely American Gospel, one we sorely need to recover.

What follows is a parallel narrative—a personal history of how I slowly came to discover and understand this gospel, and a history of how that gospel arrived and evolved in this country. From the founding of Jamestown four hundred years ago, up through the founding of America's only homegrown philosophy, pragmatism, this other gospel has been reappearing, reinvented, again and again in the writings of certain American geniuses, thinkers like William Byrd, Thomas Jefferson, Ralph Waldo Emerson, Walt Whitman, William James, John Dewey, and Lynn Margulis. Together, they offer an astonishingly comprehensive and relevant vision for where, and how, the United States must proceed in the twenty-first century. I believe these men and women can show us a way back to our country's best impulses, and thus a way forward to a future that is more respectable, more responsible, more sustainable, more interesting, more reverent.

I

First Shot

I have only one photograph of my father. It's a forty-year-old Polaroid that my mother found in a shoe box a couple of years ago when she was cleaning out her closet. She took the picture on Christmas Day, 1967, when I was seven months old. I know this because I am standing beside my father in the picture. Or not exactly standing. My father is bent over, holding me upright with one hand. He is wearing black dress shoes, gray dress pants, and a white button-down shirt. He looks serious, overly serious, and uncomfortable. He looks for all the world like a young seminarian, which, a year earlier, he had been. In this picture, we are posing in the living room of the parsonage of his first church. Cheap paneling covers the walls; cheap linoleum covers the floors. My father is wearing black horn-rim glasses, and a pipe juts out from the right corner of his unsmiling mouth.

This is the reason for the photograph. Standing knee-high to my father in my pajamas, I am holding a pipe up to the right corner of my mouth, though it appears I am trying to smile. That pipe would have been a present from my mother, just opened by my father. I must have picked it up and put the stem in my mouth, and that must have inspired my mother to capture this image of father and son, fitted out with their pipes.

At our feet is a copy of *The Washington Post*. And though the colors of this snapshot are fading, I can still make out an image of Lyndon Johnson there on the front page, above the fold. After the Gulf of Tonkin Resolution, my mother's twin brother had decided it was time for him and his new wife to start a family. The birth of my cousin Cindy in 1966 won my uncle a deferment from Vietnam. My own father, being a minister, had already been granted 4-D status, which exempted the clergy from service. He would die in another war, one fought with the hunting rifle that stands behind the open door that leads us out of this picture.

Open Eyes

For sixty years, my grandfather was a country preacher in the Tidewater area of eastern Virginia. He used to tell me I was named after the Viking pirate Erik the Red, who was kicked out of the Scandinavian colony on Iceland by his own people around the year 980. After that, Red Erik packed up his family and sailed to Greenland to start a new colony. His son Leif later made the two-thousand-mile voyage east and landed on Newfoundland five hundred years before Columbus set sail. A statue of Leif Eriksson now stands outside the Mariners' Museum in Newport News, Virginia. When I came to visit my father's parents in the summer, my grandfather used to take me to see the bronze Viking staring out across the horizon. Back then, we did many maritime things together. Each year, my grandfather would announce from his pulpit that I had arrived for my summer visit, and the men in the congregation would put their boats at our service. We fished the Chesapeake Bay, sailed along the James River, canoed the inlets around my grandparents' riverside home, and hunted for sharks' teeth everywhere we went. With a piece of string tied around a raw chicken neck, we would lure crabs into a dip net below my grandparents' dock, which jutted into a cove beside their house. When a crab refused to untangle itself from the strings of the net, my grandfather would smack it with a sawed-off baseball bat, and the startled crab would fall into our galvanized pot.

Because I spent much of my childhood with my grandparents, the early history of the British presence in the New World—most of which happened in Virginia, after all—was very real to me. I knew that Virginia had been named for the Virgin Queen, Elizabeth I, and that back then, its perimeters stretched north past the Great Lakes. Throughout Colonial Williamsburg, I could show you which houses still had Revolutionary War cannonballs lodged between the bricks and mortar. Or at the College of William & Mary, I could point out how, over the years, fires had badly warped the three-foot-thick walls in the classroom buildings. And above all, I hated the fat British general Cornwallis as if he had personally tried to deprive my own family of life, liberty, and the pursuit of happiness.

But my favorite part of the Tidewater area of Virginia was Jamestown—or at least the historic re-creation of the first British colony in the New World. I was fascinated by the three ships docked there, replicas of the *Discovery*, the *Godspeed*, and the *Susan Constant*. The boats seemed so small to have carried 120 settlers to North America in 1607. Later I would learn that conditions on the ships had been disgusting, that Captain John Smith had managed to make everyone's life more miserable, and that, according to the Reverend Robert Hunt, passengers regularly "made wild vomits into black night." But the colonists were determined to find timber with which to expand the British navy, and the Church of England did not want to forfeit the New World to Spanish Catholics. So Britain's poet laureate Michael Drayton lauded Virginia as "earth's only Paradise."

When the first fleet did sail into Chesapeake Bay, Captain George Percy confirmed this assessment: "I was almost ravished at the first sight thereof." And when he encountered some of the indigenous people, Percy termed them "[as] goodly men as I have seen, of savages or Christians." But the cavalier gentlemen who made up most of the 1607 crew— Smith called them "tuftaffaty humorists"—were more interested in finding gold than in growing corn, if they had to work at all. It seemed much easier to barter beads and trinkets for Indian corn. That first winter, the powerful chief Powhatan supplied the English with food, but two years later, things had soured between the colonists and Powhatan. He had grown resentful of their cannons and Smith's refusal to trade guns for food. "Many do inform me," Powhatan told Smith, "your coming hither is not for trade, but to invade my people, and possesse my country." It probably didn't help that Smith, in an effort to impress the natives with a show of strength, was sailing up and down the eastern coast in the *Discovery*, setting Indian homes on fire.

So when the Virginia Company called Smith back to England, Powhatan resolved to let the colonists starve by ambushing any who left Jamestown in search of food. It wasn't pretty. The sixty colonists who still survived were digging up corpses for food. In *Love and Hate in Jamestown*, David A. Price recounts the grisly story of one British gentleman who murdered his pregnant wife while she slept, "chopped apart her remains, salted them, and feasted on them."

Not until Thomas West, a.k.a. Lord de la Warr, arrived with supplies in 1610 and established a strict regimen of work and churchgoing did Jamestown finally get on its feet. Soon after, John Rolfe planted in Jamestown a variety of tobacco that the English craved. The Virginia Company would finally realize a return on its investment—not in North American gold, but in its gold leaf.

Two years later, the mariner Samuel Argall learned that Powhatan's favorite daughter was visiting the tribe of King Patawomeck. Argall sailed straight to the Patawomeck village of Passapatanzy and kidnapped Pocahontas. On hearing of the capture, her father immediately released seven English prisoners and sent to Jamestown a canoe filled with corn and the tools his warriors had stolen from the colony. The British demanded that stolen guns be returned as well. Powhatan replied that he had no guns; they were either broken or had been stolen from him. During a standoff that lasted for months, Pocahontas was taken to the Henricus settlement to live with a minister's family so she could learn English, and thereafter, the teachings of Christianity.

In March 1614, Lord de la Warr's ill-tempered deputy, Thomas Dale, decided to force matters to a head. He sailed with 150 men and Pocahontas to one of Powhatan's villages on the Pamunkey River. Powhatan himself was several days' journey away, but many of Pocahontas's half brothers were there. She stepped from the ship and announced coolly that if her father would not trade guns and old axes for his own daughter, then she would just as soon stay with the English. During her months in Henricus, it seems, she had become captivated with the Christian faith, as well as with the tobacco planter John Rolfe. While the ship was moored at the Powhatan village, Rolfe asked Dale for permission to marry Pocahontas. Though Dale thought the natives inferior in all ways to the English, with his calculating nature he must have immediately seen that this union might mean a new peace with Powhatan's people.

He was right. Rolfe and Pocahontas were married the next month, and the new bride became the first native of eastern North America to convert to Christianity. A period of relative peace ensued between the settlers and Powhatan's people.

The myths and history of Jamestown are of course Hollywood fare by now (though no one ever seems to give Rolfe credit as the Indian princess's true love). But the thing I always liked best about the re-created colony was its potter. On days when my grandfather was working on his sermon or visiting hospitals, my grandmother would wait on a shaded bench while I watched this man, dressed in a white period shirt with baggy sleeves, make one pitcher after another magically rise from a lump of clay. Sometimes I would actually see him gathering a large basket of clay on the banks of the James. Back in his shop, he would pound out air bubbles from the clay, then, with a piece of wire fastened to his workbench, cut large slabs of that raw material in half, over and over, until it was finally ready to be spun and shaped into a bowl or a pitcher. I remember he had curly hair and always seemed completely absorbed in the task at hand. He never spoke, and somehow, that made his work seem to me all the more important, even reverent. Though I knew I was supposed to grow up to be a preacher like my father and grandfather, I secretly wanted to be a potter. To spend all day making things—beautiful, simple things that people needed and would pay for—that kind of life appealed to my eight-year-old self in a way the ministry never would.

Though my grandfather used to tell me I was a descendant of Red Erik, in truth we never traced our family lineage back beyond a Norwegian woman named Wilhelmenia Frederickson. She was my grandfather's grandmother, who married a Swede named Will Buhler and then sailed for Chicago in the 1880s, nine hundred years after Red Erik's maiden voyage. For some reason, when the couple was being processed through Immigration, they were given the name Anderson (common enough for Scandinavians). My great-grandfather Harry Anderson was born in 1887, shortly before his father collapsed from a heart attack. Left with six children, Wilhelmenia went to work scrubbing floors at night in Chicago office buildings.

At age five, my great-grandfather Harry went to work, thinning and harvesting in a large Chicago onion field. With a small wagon, he also hauled laundry from wealthy families

to his mother, who washed it all on a scrub board, then sent
Harry back with the clean, ironed clothes. His oldest sister,
Christina, began volunteering for the Salvation Army. Metal
pail in hand, she would solicit pennies from men in taverns
for the greater good of the Temperance movement. According to my grandfather, when Harry was older, he would mock
his sister's piety, imitating her in a high-toned voice: "I am *in*
this world, but not *of* it."

Harry eventually started a business manufacturing window blinds and in 1920 married Mabel Clark. They bought
five rural acres outside Chicago. By himself, he built a small
house, set the chimney, laid the wiring, and dug a well. He
built chicken coops and planted an orchard. Still, by all accounts, my great-grandmother was never quite satisfied. She
wanted something nicer, something closer to Lake Michigan.
She didn't want chickens in her yard. She especially disliked
it when Wilhelmenia came to visit and got down on her knees
to scrub the floors.

As a result, Harry saw less and less of his mother. During
the Depression, he came to believe, somehow, that she had
gone to live with his sister Christina. But one winter morning,
Wilhelmenia was found frozen to death in a cardboard
house, set above a warm air duct on a Chicago sidewalk. "I
remember Dad crying so hard at her funeral," my grandfather's sister Rosemary wrote him in a letter, now tucked

inside the family genealogy book, which has, more or less by attrition, come down to me.

At that time, my grandfather David Anderson was away at a small Bible college. When he graduated, he married another student, Antoinette Johnson, and was soon asked to pastor a church on the banks of the James River in Tidewater, Virginia. My grandfather spent the rest of his life preaching to fishermen and their families. He preached a simple theology to hardworking people who had little need for scriptural ambiguity or liberal skepticism about the Bible's inerrancy. This was the word of God, plain and simple, and my grandfather was handing it down. He was never a narrow-minded man, but he was a charismatic, fundamentalist preacher. His faith was the center of everything he did. He was the kind of man who would begin the Sunday blessing and reduce himself to tears as he contemplated in prayer Christ's love of mankind. He was slightly built, with strong legs and narrow shoulders. Heavy bags always hung beneath his eyes. He spoke softly and moved with a quiet, serious bearing. But on Sundays, in a white polyester suit and a light-blue shirt, he exploded in front of the church, often bringing his fist down hard against the pulpit to emphasize the severity of the sacrifice Jesus Christ had made for us all. He clarified and amplified the Bible for his congregations, and they loved him for it.

———

I have an old Memorex cassette, given to me by my grandmother, that is a recording of my grandfather preaching her favorite sermon. It is called "Open Eyes," and it takes as its central text the first three books of Genesis. The tape begins with the choir of the Union Baptist Church in Achilles, Virginia, accompanied by my grandmother on piano, singing:

How I love you, blessed Jesus,
you are more than life to me;
all my sins have been forgiven,
by your death on Calvary.

Then my grandfather steps to the pulpit. I can hear him clear his throat and in my mind can see him take off his wristwatch and place it beside his Bible.

"There are many passages in the word of God where it says God opened people's eyes," he begins. Today, he tells the congregation, he wants to talk about a conversation between Eve "and the adversary of our souls, the devil." The "hissing stinger," as my grandfather translates the Hebrew, was a beautiful animal, and "the devil, Satan, entered into this beautiful animal, the serpent."

Then, as an aside, my grandfather says, "It must have been

a very significant time in the human race when animals actually talked to people." There seems to be something almost incredulous in his voice, as if he might not really be the literal-minded fundamentalist we all take him to be, as if he might even allow for a figurative element within this story. But as was usually the case with my grandfather, one could never be sure.

He goes on, telling the story of the serpent's tempting promise of knowledge, of Eve eating the fruit and then coaxing Adam into the transgression of the Creator's one prohibition. But my grandfather's focus is on the serpent's prophecy: "For in the day you eat from the tree, your eyes shall be opened." Then he clarifies the serpent's promise: "You are going to have something God doesn't want you to have."

My grandfather pauses. His face and neck would have reddened by this point; his carotid artery would be visibly pulsing. Then he drops the hammer: "And you know, dear friends, that has been the problem with the human race ever since the very, very beginning. God is the Creator, and He has the authority of the potter over the clay. But man does not want to admit that he is merely clay. God, in creating us, also has the right to demand of us that which he prescribes. And God has taught us very clearly that, as Creator, he gives to us the awesome power of free will. We have the instincts of animals in some respects in our bodies, but we have the minds and the hearts of God Almighty. He has made us rational beings. And he has told us that there is the basic law of the harvest in this creation: whatsoever a man soweth, that shall he also reap. He that soweth to the flesh, says Galatians, shall of the flesh reap corruption. But he that soweth of the spirit shall reap life everlasting.

"From the beginning of time, from the creation of the human race, the devil has been lying to the human race. And instead of the human race seeking the truth, which is found in obedience to the revealed word and will of God, the human race has allowed itself to be led astray, even as Eve was in the beginning. Her eyes were opened, and the moment her eyes were opened, she became guilty. Friends, you can't cover your guilt. You can't cover your guilt with a fig leaf. The only covering for the guilt of human beings is the precious blood of Jesus Christ. It's the only substance in all the universe that can cleanse a man's heart of guilt and stain."

My grandfather has a few more things to say, but what Baptists call "the invitation" is in sight—sinners would soon be invited to kneel beside my grandfather at the front of the church and ask Jesus to cleanse the stains from their hearts. "My dear friends, beware, beware, of those who say, 'Listen to me, and I will open your eyes.' Go to the word of God, ask for the enlightenment of the Holy Spirit, and say, 'O Lord, in the spirit of humility, I come to thee. On the basis of the shed blood of Jesus Christ I come to thee.'"

In the few seconds of silence that follow on the tape, I can imagine my grandmother rising to take her place back behind the piano, and then I hear her start to play the invitation hymn, "Break Thou the Bread of Life."

Bread. Flesh. Sin.

Guilt. Blood. Sacrifice.

Accommodation

My father, like his father, was a Baptist minister. He shot himself with his hunting rifle when I was three. Twenty years later, my mother gave me his Bible, the one he had used since his days in seminary. I opened it to the cloth marker at chapter 10 of Matthew, and read:

Never imagine I have come to bring peace on earth;
I have not come to bring peace but a sword.
I have come to set a man against his father,
a daughter against her mother,
a daughter-in-law against her mother-in-law;
yes, a man's own household will be his enemies.

He who loves father or mother more than me is not
  worthy of me;
he who loves son or daughter more than me is not
  worthy of me:
he who will not take his cross and follow after me is
  not worthy of me.
He who has found his life will lose it,
and he who loses his life for my sake will find it.

(Matthew 10:34–39, trans. James Moffatt)

I do not know—have no way of knowing—if the marker had fallen on that page simply by coincidence, or if it reveals something crucial about what drove my father to suicide. But the fact that the question, like all the questions we ask of suicides, is unanswerable makes the second possibility inevitably more compelling, and more sinister. Who is the egomaniac speaking these words? Who would demand across centuries that we love him, a historical stranger, more than our own fathers, mothers, sons, and daughters? And just a few pages earlier, wasn't this same teacher bestowing blessings on the peacemakers? Given such a wild reversal, we might, with good reason, doubt the authenticity of the passage. But what I want to know, and have no way of knowing, is whether my father took this passage seriously.

From everything my mother, my father's friends, even my

grandfather, have told me, my father wasn't the kind of man who would have read these lines literally. Much of the anxiety he began to feel as he grew out of adolescence was related to the fact, and the fear, that a gap was widening between his own beliefs and those of my grandfather. Because there was such a clear distinction in my grandfather's mind between good and evil, spirit and flesh, the word of the Lord and the godless static of the world, answers came easily to him. And perhaps because they did come so easily, and were greeted so uncritically by his congregation, I don't think it ever occurred to my grandfather that he might be wrong about anything.

Such certainty lent my grandfather his strength of conviction, his power before crowds, but I also think it came at the cost of self-awareness. I sometimes suspect this is a trait common in Baptist ministers, but it isn't one my own father possessed. When he was pastoring his first church, a woman from his congregation would come by the parsonage every Sunday after church to give pointers on how my father might improve his sermons. My mother told me he would always listen patiently and then grow quiet and sullen after she left. I try to imagine the same woman interrupting my grandfather's Sunday dinner. He would have dispatched her with such righteous haste that his butter beans would have still been steaming when he returned to the table. But my father lacked that conviction, and therefore lacked the confidence that comes with it. If my grandfather had too little self-awareness, my father seemed to have too much. It led to a debilitating self-consciousness and a depression so heavy he could not rise out of bed some days. He was in such a state when my mother left to teach school on that last morning. In the afternoon, she collected me from the babysitter's and returned to our yellow clapboard parsonage in Wakefield, Virginia, to find my father still in bed. But this time his body was lifeless. His rifle lay at his side.

That was September 1, 1970. Fourteen years earlier, in the summer before his freshman year of college, my father swallowed a bottle of aspirin. He told my grandfather that he was confused. Besides, when they had been on vacation in Canada that summer, he overheard teenagers making fun of his clothes. "And you expect me to go to college wearing these baggy pants?" he told my grandfather, who took him down to Knockman's department store the next day. My grandfather told the saleswoman, whom he knew, "I've been blind here. I've been expecting Don to go to school wearing these baggy pants, and evidently it's become a matter of real concern." He was trying to make a joke out of it. She played along, measured my father for new pants, then said, "Okay, then, Don, I pronounce you fit for school."

But about a month later, my grandparents received a call from Carson Newman College; they were told my father wasn't getting out of bed to go to class. The school suggested my grandparents come down and collect their son before the six-week deadline, when a "withdrawal" would appear on his official transcript. So that's what they did. Driving back to Virginia, my grandfather was still at a loss. "I thought we had you all fixed up," he told my father. "I thought we had you looking like somebody, not like some hick son of a country preacher."

But my grandfather must have known by then that my father's problems went much deeper than the sartorial. My father had always been oversensitive to the culpability that loomed large in my grandparents' religion—the belief that Jesus died as a sacrifice to cleanse a fallen world of its sins. And for Southern Baptists, particularly those of my grandparents' generation, sin is everywhere. As the twentieth century became rapidly, increasingly, secular, the spiritual life must have gradually begun to seem to my father like an ever-lengthening list of things that must be avoided: women, alcohol, playing cards, foul language, movies.

My mother grew up in a small town near Macon, Georgia, and met my father when they were both doing summer missionary work in Montana in 1965. My mother was as voluble and sure of herself as my father was reticent and awkward.

And whereas Christianity was beginning to seem to my father like a penal system, to my mother, it meant just the opposite: absolute liberation. Jesus was the truth that set us free. That was, and is, the heart of her religion. It must have looked very appealing to my father; surely he could have used some of that lightness. In the end it wasn't enough, of course, and my grandparents found subtle ways to let my mother know she had failed my father. Though his life insurance policy barely paid for the funeral, they never offered to help my mother with the costs of raising a three-year-old on her own.

My mother taught high school for three years; then, on another mission trip, she met Tom Reece, the man who would become my stepfather and eventually adopt me. They married, and we moved to Louisville, Kentucky, where my step-father worked as a mechanical engineer. My mother and stepfather did not raise me in the atmosphere of overbearing piety that suffocated my father, but their beliefs were fundamentally the same as my grandparents'. Heaven and hell were real, and there were consequences for misbehavior. I had to be at church every Sunday morning and Wednesday night, I had to sing in the choir, and above all I had to conduct my-self in a way that would not disappoint my grandparents.

To my mother's horror, somewhere along the line I inherited my father's dim view that Christianity was a set of rules meant to inflict self-loathing. In trying to please my grandparents, I too had come to find that *goodness* can become paralyzing and impossible. How could I think this? my mother wanted to know. Had she ever said that? Didn't I know that Jesus *freed* us of our sins? No, she concluded, it must be my grandparents' fault. Somehow, I must have formed this warped notion during all those summer vacations spent in Virginia.

It was the summer after my senior year in high school when things really began to unravel. Our church burned down, and that seemed to me as good an omen as any that I had no more use for its choir loft, its Sunday-school rooms; I could even do without its indoor basketball court. My parents decided to join a new church, and I refused to go along. Then one night I made it known in an ill-timed comment over dinner that I thought prayer was a waste of time. Everyone began shouting at once. It was the only time I seriously thought my stepfather and I would come to blows. When I left for college, everyone was relieved.

In fairness, I cannot lay all of the blame for my father's suicide at the steps of the church. He suffered from bipolar disorder, as did his younger brother, whose life, up to the day he died, remained a haunted search for stable work and the right medication. My father hoped marriage and family would cure him. So he hid his problems from my mother until they were married, until he couldn't get out of bed. Rounds of therapy followed, then increased dosages of vitamins, then rounds of shock treatment. Nothing helped, and in the end, a .22 gauge Remington looked to my father like the only solution. "I honestly think he felt like he was doing us a favor," my mother says today without bitterness, "like we would be better off."

My grandmother always found consolation in the fact that my father was thirty-three—the "same age as Christ," she would say—when he died. Several years ago, when I turned thirty-three, I started to experience incredible migraines for the first time in my life. Then a numbness started to creep up and down my arms and legs. No one could explain it. The CT scans showed no signs of the tumor I feared; the MRI showed no signs of multiple sclerosis. But the symptoms remained. Then my skin began to crawl. No amount of Gold Bond lotion could stop or sooth the itching. Panic attacks began. Doctors tried all manner of pharmacological relief. Nothing helped. Thirty-three was the worst year of my life.

Finally I sought out a therapist. I told her my story. I told her I was the same age my father was when he killed himself. After several sessions, she told me I needed to have a serious talk with my grandfather. Over Christmas, I drove down to my grandparents' retirement community in Florida. During the last few years, my grandmother had quickly begun losing her short-term memory, and my grandfather was having one heart attack after another. Because he could no longer take care of my grandmother, he had moved her to an assisted-living facility. When I visited with my grandmother in her new room, I did all of the talking, mostly about the books she had read to me when I was a kid. As I talked, her mouth would drop open, her chin would quiver, and she seemed to be staring hard at the ceiling, as if that were where her memories of all these events had disappeared.

On my last night, my grandfather and I sat down at his small kitchen table, and with a tape recorder running, I asked the questions I had never asked about my father's suicide. That's when my grandfather told me about the bottle of aspirin, the baggy pants, the hunting rifle he had given my father. He said it was true that the church, religious authority, and his own fundamentalism had led my father to a crisis he could never resolve. "I think your father was confusing what God wanted him to do and maybe what I and his mother wanted him to do," he admitted. "We had to finally realize that there were some questions in Don's mind between religious authority and human freedom. As time went by, I could see that he felt he was not free." This is what I had expected, had wanted to hear. But then my grandfather said something

I didn't expect. He said that he himself had thought about suicide hundreds of times. But, he said, "I do not have the character, the courage, to do what your father did. To take my own life." The character? The courage?

"The poor child felt that he would never be able to accommodate himself," my grandfather said. And that, I realized, was exactly it. To *accommodate*: to find some middle place between where one comes from and where one hopes to arrive. Or: to pretend one is at home when, and where, one isn't. That is what my father could never do. He could never find accommodation—either within his father's religion or with the treatment of his own depression. But suicide is the one human act that requires no accommodation. The suicide need never return, like the prodigal son, to the father's house. He is unbeholden. The suicide cannot be questioned; the suicide cannot be punished. To raise the sword against oneself is to find release into a moment—an endless moment—of freedom.

William Byrd's New World

Eighteen years of compulsory churchgoing, followed by eighteen more years of trying to extract myself from the church,
has convinced me that Tolstoy was right: the more Christian
fundamentalism emphasizes that the kingdom of God awaits
as a reward in the afterlife, the more it ignores the teachings
of the Sermon on the Mount. That is to say, the more mainstream Christianity emphasizes a theology of salvation from
this world, the more it ignores Jesus' teachings of how we
should act while we inhabit this earthly realm. And the
more Christianity becomes about salvation from this world,
the more it highlights our inherent sin, the sole obstacle to
salvation. The result is what the American philosopher William James, to whom I will return more than once in this

book, called "the sick soul." Clearly, my father suffered from such a sickness. And its source, if Puritan literature is any indication, extends back to the Massachusetts Pilgrims. The colonial poet Edward Taylor described himself as "a varnisht pot of putrid excrements," while Cotton Mather wrote in his diary that he was a "poor, sorry dispicable Vessel . . . fit for nothing but the Dunghill." For these men, identity seemed defined by nothing so much as scatological excess. And in his famous sermon "Sinners in the Hands of an Angry God," Jonathan Edwards stormed:

> There is laid in the very nature of carnal man the foundation for torments in hell. Your wickedness makes you, as it were, heavy as lead and to tend downwards with great weight and pressure towards hell. . . . The God that holds you over the pit of hell, much as one holds a spider, or some loathsome insect, abhors you and is dreadfully provoked; His wrath towards you burns like fire; He looks upon you as worthy of nothing else but to be cast into the fire.

After Edwards delivered that sermon during a Northampton revival in 1750, a man left the service and hanged himself. The 220 years between that suicide and my father's suggest how persistent and influential Puritan fundamentalism has remained in this country.

In his experimental, alternative history *In the American Grain*, the modernist poet William Carlos Williams tells of attending a Paris dinner party in 1924 where he fell into a heated conversation over the virtues of the Puritan minister Cotton Mather, and of Puritanism in general as a source of American culture. Williams's French interlocutor found Mather's sermons and writings "a little grotesque," but "firm" and full of "vigor." Williams took exception. The fixed religious doctrines of the Puritans left no room for curiosity about either the New World or its indigenous people. William Bradford, the first governor of Plymouth Plantation, did not rejoice, as had George Percy when he set foot on the continent. Whereas Percy had seen a New Eden, Bradford saw "a hideous and desolate wilderness." And whereas Percy found noble savages lurking at the edge of the woods, the Puritans uniformly found devils. In 1634, the English minister Joseph Mede put forth the theory that just as the Puritans were the New Israelites led as God's chosen people to Plymouth Plantation, so the devil had led the Indian heathens into *his* own chosen land—the wilderness of North America. However, William Carlos Williams believed this self-anointed purity "bred brutality, inhumanity, cruel amputations." And it's true: the more the Massachusetts Pilgrims became obsessed with religious purity, the more barbarous their behavior became. The very first lines of Increase Mather's "A Brief History of the War with the Indians in New-England" betray the arrogance of those who think themselves the pure, elected people of God:

> That the Heathen People amongst who we live, and whose Land the Lord God of our Fathers hath given to us for a rightfull Possession, have at sundry times been plotting mischievous devices against that part of the English Israel which is seated in these goings down of the Sun, no man that is an Inhabitant of any considerable standing, can be ignorant.

When the Pequot Indians showed the temerity to try and take some of their land back during what became known as King Philip's War, they were ruthlessly slaughtered. When Captain John Mason's men set fire to a Pequot village on the Mystic River, a community of mostly women and children, Increase Mather's son Cotton wrote, "It was supposed that no less than 600 Pequot souls were brought down to hell that day."

What comes through most forcefully in these early American histories is a millennial arrogance: we will build the City on the Hill, which will usher in the Second Coming, when

God redeems the Elect Few. The Puritan legacy has remained an influential part of our collective American psyche, and there are a few important conclusions to be drawn from this. The Puritans' emphasis on the End Time led to their obsession with salvation *from* this earth. This led not only to a hostile view of the natural world but also to a hostile view of its inhabitants, who practiced their own religion. "Preach much about the misery of the state of nature," Charles Chauncy urged fellow Congregationalist ministers. What's more, the individual's anxiety about whether or not he or she was one of the Elect led to a crippling desire to constantly purge one's soul of sin. The more the Puritan preachers emphasized the inherent sinfulness of human beings, the more the early Americans came to feel a certain self-loathing. The more they came to hate themselves, the more hateful and violent they became toward those (Quakers, women, indigenous people) they perceived as impure. James Baldwin understood perfectly the psychology behind such violence. In *The Fire Next Time*, he argued, "White people in this country will have quite enough to do in learning how to accept and love themselves and each other, and when they have achieved this . . . the Negro problem will no longer exist." Though Baldwin doesn't say so directly, the self-hatred at the heart of Puritanism may explain much about white America's intolerance of the Other. The self-denial of one's own nature quickly became synonymous with a denial of all nature, and the lasting effects of this philosophy would prove perilous for the New World.

In the end, William Carlos Williams told his French counterpart that he, Williams the American writer, had to "disentangle the obscurities" of Puritanism "that oppress me, to track them to the root and to uproot them." The metaphor is apt. Puritanism was always an invasive species in this country, an unnatural migrant that, as Williams said, needed to be uprooted. It was an indoor religion that understood nothing of what Henry David Thoreau would later call the "unroofed book" of nature. But there is a more truly American literature and a more truly American religion rooted in the American landscape.

It began, not with the sick soul, but with its opposite— what William James referred to as *the religion of healthy-mindedness.* And I believe it began at the turn of the eighteenth century, with a nearly forgotten Virginian named William Byrd. Unlike Increase and Cotton Mather, Byrd was one of those free spirits who, as James wrote, "seem to have started in life with a bottle or two of champagne inscribed to [them]." He was a man utterly fascinated with nature and human nature. Whereas the Puritans saw in nature a fallen world, Byrd saw a depository of the holy. His book *The History of the*

Dividing Line, about a surveying trip along the Virginia–North Carolina border, would become, according to historian Roderick Frazier Nash, "the first extensive American commentary on wilderness that reveals a feeling other than hostility." Thus Byrd wrote the first chapter of a free-spirited American gospel that would abandon the sick soul for a religion that was joyous, sensual, reverent, earthly.

With his wide-ranging interests and intellect, William Byrd appears in retrospect as something of a Thomas Jefferson before Thomas Jefferson, a Leonardo of the New World. Byrd served as a member of the Virginia House of Burgesses. He had been raised on his father's 2,600-acre plantation called Westover, on the banks of the James River. He was sent to London in 1684 to study law. When his father died in 1705, Byrd had just been elected to England's preeminent scientific institution, the Royal Society. Like Jefferson, who would later attempt to have *The History of the Dividing Line* published, Byrd considered himself a man of the arts *and* sciences. But he returned home to manage the plantation and take his father's place as auditor of Virginia. He expanded Westover's gardens and orchards and built a handsome Anglican chapel, which he attended every Sunday. He rose each morning at five to pray and read Greek and Hebrew. His library contained more than three thousand volumes, and the walls of Westover were covered with an expansive art collection. Byrd gradually added 2,900 acres to the plantation, and he never tired of walking those vast tracts, inspecting plants and collecting the herbs he would, as an inveterate naturopath, turn into potions for sick friends. In the words of Louis B. Wright, who edited his collected prose, "Byrd was never bored."

Like Thoreau years later, Byrd believed the elemental act of walking to be almost a spiritual exercise. Even when Byrd's party was trudging through North Carolina's malarial Dismal Swamp, Byrd reported, "this dirty Dismal is in many parts of it very pleasant to the eye." He chided his fellow Englishmen for their timidity in regard to the American landscape. "Our Country has been inhabited more than 130 years by the English," he wrote, "and still we hardly know any thing of the Appalachian Mountains, that are no where above 250 miles from the sea. Whereas the French, who are late comers, have rang'd from Quebec Southward as far as the Mouth of the Mississippi, in the bay of Mexico, and to the West almost as far as California, which is either way about 2000 miles." While the Puritans retreated to their City on a Hill in fear of the wilderness and its inhabitants, Byrd found the natural world spectacular in both its complexity and its beauty.

What's more, he was, again like Thomas Jefferson after him, an adamant defender of Virginia. He began *The History of the Dividing Line* by "shewing how the other British Colonies on the Main have, one after the other, been carved out of Virginia." Byrd clearly took this as an affront, nor did he hide his suspicion that, had North America remained under Virginia's auspices, many "Wars and Massacres," not to mention witch trials, could have been avoided. The one splinter group for which Byrd did hold out praise was the Quakers. Flogged, mutilated, and executed by the Massachusetts Puritans, the Quakers moved to Pennsylvania, where "they have the Prudence to treat [natives] kindly upon all Occasions," and so lived in peace with them. It was a policy Byrd himself strongly advocated.

For decades throughout the late 1600s and early 1700s, the boundary line between Virginia and North Carolina had been a growing source of aggravation on both sides. The Virginia House of Burgesses resolved to send parties from both colonies to survey and establish an official boundary line. In 1727, on his fifty-third birthday, Byrd was named by the recently crowned George II to head up a Virginia delegation. On March 5, 1728, the commissioners, surveyors, laborers from North Carolina and Virginia, along with a chaplain, departed from Currituck Inlet, North Carolina, with tents, bags of flour, and, by Byrd's assessment, "as much wine and rum as will enable us and our men to drink every night to the success of the following day."

In *The History of the Dividing Line*, Byrd neglected neither the "limpid and murmuring" streams lined with "Myrtles & Bay-Trees" nor—in his private version, *A Secret History of the Dividing Line*—the various complexions of the young women he encountered along the way. And Byrd's prose style is as graceful and witty as Cotton Mather's is tortured and dry. Of one innkeeper's daughter, Byrd wrote, "She was a smart Lass, & when I desired the Parson to make a Memorandum of his Christenings, that we might keep an Account of the good we did, she ask't very pertly, who was to keep an account of the Evil. I told her she shou'd be my Secretary for that, if she wou'd go along with me."

Indeed, if the private diaries Byrd kept while earning his law degree in London are any indication, it appears the innkeeper's daughter would have been quite busy. Like his British playwright friend William Wycherly, Byrd was a bit of a rake. He wrote in his diary on August 24, 1719, "Then I went to visit Mrs. A—l—n and committed uncleanness with the maid because the mistress was not home. However, when the mistress came I rogered her and about 12 o'clock went home and ate a plum cake for supper. I neglected my prayers, for which God forgave me." (After he was married, Byrd wrote in his journal that, after a spat, he preferred to reconcile with his wife "on the billiard table.") Unlike Mather, Byrd believed in a very forgiving God.

Today, Dover Books publishes *The History of the Dividing*

Line and *The Secret History of the Dividing Line* in a helpful edition where the accounts appear side by side to highlight their differences. The first *History* is the report that Byrd presented to the Lords of Trade to ensure that he and his men would be properly compensated. It is a detailed description of the natural world, the geography of the lower Appalachians, and the customs of its inhabitants, both native and European. *The Secret History*, written only to amuse Byrd's friends, is an account of the same trip that reads more like a rollicking eighteenth-century British novel, full of intrigue and grab-ass. It is a Swiftian satire of human vanities. In it, all the significant members of Byrd's entourage are given symbolic monikers such as Astrolabe and Puzzlecause; the minister is deemed Reverend Humdrum. The quarrelsome Virginia commissioner Richard Fitz-William, who constantly sided with the North Carolinians over boundary disputes, earned Byrd's ire and the name Firebrand. Professor Percy G. Adams has noted that at numerous inns and encampments, "the surveying party managed to find willing women everywhere they went—white, black and red." And when men from Virginia imposed themselves on women who were not so willing, Byrd usually blamed Fitz-William for instigating these affronts. One landlady, Byrd noted, "fortify'd her Bed chamber & defended it with a Chamber-Pot charg'd to the Brim with Female Ammunition" to fend off Fitz-William.

Or Byrd blamed bad North Carolina liquor. His elitism as part of the Virginia gentry is often on display throughout *The Histories*. As his biographer Richmond Croom Beatty observed, Byrd showed "an amusing disesteem of everything related to North Carolina." He was repulsed by the slovenliness of the North Carolina woodsmen he met, reasoning that "these People live so much upon Swine's flesh," they had become "extremely hoggish in their Temper." Nor could a man with so much innate curiosity comprehend the indolence of men "who make their wives rise out of their beds early in the morning, at the same time that they lie and snore till the sun has risen one-third of his course and dispersed all the unwholesome damps." Byrd found the same idleness in the native people he met, but he thought they came by it naturally (at least until the English introduced them to rum). They were expert hunters who did not possess the cravings and ambitions of the English. A century later, Thoreau would note that the Penobscot Indian owned his teepee, while every two-story house in Concord belonged to the banks.

When the surveying party finally ran out of liquor somewhere around the Roanoke River, the North Carolina contingent decided they had taken the boundary line far enough. Fitz-William, as usual, agreed with them. Byrd said his commission was to take the line to the foothills of the Appalachian Mountains. After a confrontation that nearly ended in a duel, Byrd sent Fitz-William packing with the North Carolina surveyors, but not before the North Carolinians "were so kind to us as to drink our good Journey to the Mountains in the last Bottle we had left." The Virginians forged ahead. Byrd wisely hired an Indian guide who in *The Secret History* is dubbed Ned Bearskin. The native woodsman proved an excellent hunter, so good in fact that Byrd was able to test his theory that bear meat made men more virile; within a year of his party's return to Virginia, most of the men, including Byrd, were new fathers.

In the evening around the campfire, Byrd would press Ned Bearskin about the religion and customs of his people. Whereas Cotton Mather loathed the native "savages," Byrd found them, and specifically their religion, fascinating. He quickly learned an expedient lesson in native beliefs when Bearskin refused to produce a wild turkey he had shot, "lest we shou'd continue to provoke the Guardians of the Forrest, by cooking the Beasts of the Field and the Birds of the Air together in one vessel." When Byrd amended that kosher taboo, Bearskin provided a detailed account of an afterlife where departed souls travel either a level road to an abundant hunting ground where corn springs spontaneously from the soil, or a mountainous path leading to a desolate wasteland where the women have claws for hands. In this realm, the ill-fated souls are tortured by a "dreadful old

Woman on a monstrous Toad-Stool" until they are ready to be driven back into the earthly world to mend their ways in hopes that in the next afterlife, they will walk the level path. Byrd, for his part, was satisfied that Bearskin's theology contained "the three Great Articles of Natural Religion: The Belief of a God; The moral Distinction betwixt Good and Evil; and the Expectation of Rewards and Punishments in Another World." Which led him to conclude: "All Nations of men have the same Natural Dignity, and we all know that very bright Talents may be lodg'd under a very dark Skin. The principal Difference between one People and another proceeds only from the Different Opportunities of Improvement." It's hard to imagine a sentiment further removed from Cotton Mather's hysterical aversion to the North American natives. What's more, Byrd's solution to Indian trouble was decidedly un-Puritan. In what one might call the Pocahontas position, the sensualist Byrd advocated miscegenation, arguing that "a sprightly Lover is the most prevailing Missionary." Byrd's surveyors were happy to test this hypothesis when they stayed one night at an Indian village; the next morning, their chaplain was highly distressed to find some of their clothes covered in the bear grease that the native women used as body lotion.

Once the North Carolina contingent had departed, Byrd, no longer having to defuse rum-induced brawling, turned his attention almost exclusively to the natural world. He found rattlesnake root growing near one campsite and averred that "if you Smear your hands with the Juice of it, you may handle the Viper Safely." He had taken to chewing ginseng root throughout the day, and in the *History* he effused, "Its virtues are that it gives an uncommon warmth and vigor to the blood and frisks the spirits beyond any other cordial. It cheers the heart even of a man that has a bad wife and makes him look down with great composure on the courses of the world." He noted how the now extinct passenger pigeons "are so amazingly great, Sometimes, that they darken the Sky." As houses and inns disappeared in the more remote regions, Byrd and his men took to sleeping under the stars, "entertain'd with the Yell of a whole Family of Wolves, in which we cou'd distinguish the Treble, Tenor and Bass, very clearly." The accommodations suited Byrd's craving for wilderness. "The Truth of it is," he wrote, "we took so much pleasure in that natural kind of Lodging, that I think at the foot of the Account Mankind are great Losers by the Luxury of Feather-Beds and warm apartments." Moreover, Byrd reached the conclusion that the wildflowers themselves suffered from domestication. "I believe it may pass for a Rule in Botanicks, that where any Vegetable is planted by the hand of Nature, it has more Vertue than in Places whereto it is transplanted by the Curiosity of Man."

Byrd was one of the first Americans to suggest that we might learn more from *wildness*, the laws of nature, than from our own laws of domestication. It's an idea that has reached across hundreds of years, up to us. Wes Jackson, president of the Land Institute in Salina, Kansas, has proposed that modern agriculture should learn from, and mimic, the natural ecosystems of a particular landscape. Currently in Kansas, for instance, farmers grow annual monocultures— usually corn—that require intense use of pesticides, commercial fertilizer, and fossil fuels. But, Jackson asked himself years ago, "is it possible to build an agriculture based on the prairie as standard or model?" Such an agriculture would virtually eliminate erosion, as well as the need for pesticides and heavy, oil-driven plowing. Jackson's team at the Land Institute is at work answering the question in the affirmative, growing perennial, native polycultures in the spirit of William Byrd.

The deeper into the Appalachian forest Byrd went, the more virtue he found, until the clouds finally parted at what is now the town of Pine Ridge, and the men could see in the distance the jagged peaks of the Appalachian Mountains. Byrd was thrilled by the sight. "And that we might not be unmindful of being all along fed by Heaven in this great and Solitary Wilderness," he wrote, "we agreed to Wear in our Hats the Maosti, which is, in Indian, the Beard of the Wild Turkey-cock." Byrd wanted to push farther in hopes of finding the headwaters of the James River. However, the horses were in bad shape by then, and winter was coming on. Two hundred and forty miles from the eastern shore, the surveyors made their final blaze on a red oak, and the company turned for home. Still, Byrd felt reluctant to abandon the mountains. "We could not forbear now and then facing about to survey them, as if unwilling to part with a Prospect, which at the same time, like some Rake's, was very wild and very Agreeable." It was as if Byrd had seen something of himself in the North American broadleaf forest, something lost and vital, wild and necessary. In recording that, Byrd set in motion a particularly American strain of thought that would run through Jefferson, Emerson, Thoreau, Whitman, Aldo Leopold, Rachel Carson, Lynn Margulis, Wallace Stegner, E. O. Wilson, and Wendell Berry, among others: the natural world, far from being a hideous and desolate wilderness, is rather an antidote to the deadening forces that tried to "sivilize" Huck Finn. It is a necessary scripture, our ancestral home. *The History of the Dividing Line* marks one of the first great shifts away from the Mathers' Puritan anemia toward a much healthier spirit of belonging, both within the natural world and within one's own skin.

After *The History of the Dividing Line,* Byrd wrote two more accounts of his travels. In *Journey to the Land of Eden,* he tells of the day he "laid the foundations of two large cities"— Petersburg and Richmond. I was born in the former and moved to the latter with my mother after my father's death. But my own connection to William Byrd stretches beyond that coincidence. When I was five, two years after my father's death, my mother began dating a man whose aunt Bruce Crane-Fisher was one of Byrd's direct descendants. Mrs. Crane-Fisher still lived in the mansion that Byrd built at Westover. Modeled on the residence of the Earl of Petersborough, this Georgian mansion sits on the bank of the James River thirty miles from Richmond, where my mother taught school at the time. We would visit Westover on the weekends. My main recollection of the mansion is of always being lost in it. There appeared to be endless hallways and staircases, and it seemed I could never find my mother's room once I had wandered too far down-stairs. One night, I remember, everyone painted Easter eggs in the large green kitchen. Outside, there were stables and a large formal garden. The Anglican chapel that Byrd erected and at-tended still stood just across the front lawn. In back, a tall concrete seawall stretched down to the river. I would play in the sand there while my mother and her new boyfriend, William Byrd's blood relative, sat on the top of the wall and talked. As I found out many years later, the talk eventually turned to the emerging realization that the man might be gay, and our visits to Westover ended as abruptly as they had begun.

When Mrs. Crane-Fisher died, Westover became one more historic plantation, frozen in time and open to the public. Around that same time, while in graduate school, I first read Roderick Nash's *Wilderness and the American Mind*. When I came to the passage where Nash praises William Byrd for being the first American writer to have shown an anti-Puritan reverence for wilderness, I stopped reading and phoned my mother. Was I remembering things right? Hadn't she dated a Byrd? Was this *our* William Byrd?

It was.

"Remember the Easter egg hunt?" she gushed. "Remember how neat the garden was?"

After that, I read all of Byrd's work and tracked down whatever biographical material I could find. His presence grew large in the landscape of my imagination, a rather untrustworthy map of the actual, historical landscape of colonial Virginia.

What, I sometimes wondered, if my mother actually had married into the family of William Byrd? Would we have left behind my own family's Promethean fundamentalism for the ecumenical lightness of William Byrd, as if it were some genetic trait passed down through generations? Probably not. Still, that year of my childhood has come to signify an important distinction between the ascetic, life-negating principles of the Puritans and the beginning of a new American philosophy that found beauty, reinvigoration, even a moral law, in the natural world, and that found the human body not abhorrent but rather the subject of great curiosity, admiration, and pleasure.

The Gospel According to
Thomas Jefferson

I've spent my life moving back and forth between Virginia and Kentucky, more or less along the 38th parallel. My mother's second marriage, which changed my name from the Scandinavian "Anderson" to the Welsh "Reece," first brought me to Kentucky when I was six. A job took me back to Virginia when I was twenty-six, and when that job turned sour a few years later, I headed back to Kentucky, which had seceded from Virginia in 1789 to become the fifteenth state. Since Kentucky requires only a rear license plate, I kept my Virginia tag on the front of my truck in what I amused myself to think of as a symbolic autobiographical gesture. When

I began to study in earnest the life of Thomas Jefferson, I learned that his Kentucky–Virginia Resolution, written with James Madison, had served to render illegal John Adams's Alien and Sedition Acts. I decided that this historical turn also carried some personal symbolism for me. The Alien and Sedition Acts threatened imprisonment for any American found guilty of saying anything "false, scandalous and malicious" about the government, and even carried the death penalty for anyone, such as Jefferson, who supported the French Revolution. But for me, the Kentucky–Virginia Resolution meant that I was on the side of the First Amendment. Furthermore, I was on the side of Jefferson's self-reliant farmer, the active citizen of a decentralized government, the man and woman made virtuous by public education and religious freedom. And the other side? That was the America that had emerged while Jefferson was serving in Paris as George Washington's ambassador to France. It wasn't so much John Adams's America as it was Alexander Hamilton's. It was the America that favored a strong central government, a federal banking system, a national debt, capital concentrated in the hands of the few, and military might to protect it.

Shays' Rebellion, the popular uprising of Massachusetts farmers, demonstrates how early this country became divided into what the presidential candidate John Edwards called "the two Americas."* Ten years after the Revolutionary War, markets had grown depressed in Massachusetts, and many farmers couldn't pay their state taxes. The state legislature, dominated by the propertied upper class, refused to pass debt relief. In Worcester County, ninety-two people were jailed for debt in 1785. Farmers feared a plot on the part of wealthy Bostonians to seize their land and turn them into a tenant class. One poor farmer, Daniel Shays, who had fought alongside the French general Lafayette in the war, was forced to sell an expensive sword given to him by Lafayette so he could feed and clothe his family. Finally, in 1787, about a thousand pitchfork-wielding farmers, led by a somewhat reluctant Shays, took to the streets of Springfield, headed for the courthouse and the federal arsenal. They broke into jails

* In fact, even before General Washington had officially discharged the Continental Army, Hamilton was prodding some of the officers to support a federal tax, and a national debt, to ensure they were paid. Unbeknownst to Washington, Hamilton encouraged the armed officers to threaten mutiny if the tax wasn't passed. Washington quelled the uprising with an eloquent speech to his army. But Congress passed the federal tax, and soldiers were issued interest-bearing bonds. Trouble was, the soldiers were desperate and needed cash. So they sold their bonds at sometimes a thirtieth of their face value to Hamilton's speculator friends. Washington never fully knew the conspiratorial role Hamilton had played in manipulating the Continental Army in the name of concentrating wealth in the hands of the creditor class.

and released debtors along the way. However, by the time they reached their destination, a state militia surrounded the arsenal, and guns versus pitchforks, the militia quickly disbanded the throng of desperate farmers. Shays escaped to Vermont; fourteen other men were sentenced to death but soon pardoned out of popular sympathy for their cause.

Back in New York, Shays' Rebellion caused President Washington and Alexander Hamilton considerable consternation. It proved, to their thinking, that the Articles of Confederation were simply too weak to govern the new nation. A stronger, larger federal government was needed. Jefferson, however, responded with sympathy to Shays and his band of angry farmers. Writing to James Madison from France, Jefferson remarked, "A little rebellion now and then is a good thing; the tree of liberty must be refreshed from time to time with the blood of patriots and tyrants." It was as if Hamilton and Jefferson secretly desired some sort of role reversal: All of his life Jefferson, the slave-owning aristocrat, aligned himself with the freethinking country yeoman, while Hamilton, the son of a Scots peddler, aspired to be part of the gentry of which Jefferson claimed to want no part.

Many historians have noted how Hamilton and Jefferson competed, almost like siblings, for George Washington's approval. Both men had lost or become estranged from their fathers while they were children. Hamilton, being fourteen years younger than Jefferson, may have craved Washington's paternal affections more, and while Jefferson was away in Paris from 1784 to 1789, Hamilton established himself as the president's personal aide. He earned enough of Washington's confidence to be named the country's first secretary of the treasury. While Jefferson was getting a firsthand education on republicanism and the "Rights of Man" in Paris, Hamilton was creating a financial system that favored the wealthy, the speculators, what Jefferson called "the stock-jobbing herd." By the time he returned from France in 1790, Jefferson was astonished to find that Congress had been taken over by men who favored a British system of aristocratic governance, absent the figurehead of a king.

Hamilton was pushing through the Assumption Bill, which would allow the federal government to take over the states' debt, thereby creating a large national debt that Hamilton thought was the key to establishing the nation's credit. Jefferson was well aware that speculators were buying up nearly worthless paper money issued by the Continental Congress in anticipation of Hamilton's move. William Duer, a cousin of Hamilton's wife, was jailed in 1792 for cooking the books while at the Treasury Department, and Jefferson was convinced that Hamilton himself was guilty of speculating with Treasury funds. In effect, Jefferson believed Hamilton to be one of America's first inside traders.

Hamilton followed up the passage of the Assumption Bill with a scheme to establish a Bank of the United States that would make capital more fluid and credit more available to business interests. His political philosophy was actually quite simple. He told the Federal Convention in 1789, "All communities divide themselves into the few and the many. The first are the rich and well-born; the other the mass of the people. Turbulent and changing, they seldom judge or determine right. Give therefore to the first class a distinct, permanent share in the Government." This was precisely the kind of class-based thinking that Jefferson loathed. A national bank would give much more power to the federal government and to "the rich and well-born." It set up, as Jefferson and Madison realized, a high-stakes battle between commerce and agriculture. Both men believed that it would take power away from the agrarian states and give it to corporations. Furthermore they argued that under the Tenth Amendment, the states reserved all rights not delegated to the United States by the Constitution. Jefferson wrote to Washington, arguing that nowhere in the Constitution is the federal government given permission to establish a bank. Hamilton replied that the power to do so was "implied," and in the end, Washington decided that Hamilton had the best plan to set the nation on solid economic footing. The Bank of the United States was granted a twenty-year charter in 1791.

If before, Hamilton and Jefferson had been quarrelsome siblings, after 1791 they became irreconcilable enemies. As Joseph J. Ellis wrote in *American Sphinx*, "Hamilton was the kind of man who might have been put on earth by God to refute all the Jeffersonian values." It is remarkable that the two men most important to Washington were determined to take the country in such radically different directions. Jefferson was a defender of Thomas Paine's *Rights of Man* and the French Revolution in general. He was a republican who hated all things British. Hamilton was an Anglophile who believed that men of property should determine law. Jefferson seized any opportunity to accuse Hamilton of monarchism, even noting that at one dinner party Hamilton had proposed one toast to George Washington but three toasts to King George III. Hamilton judged the British constitution to represent "the most perfect government which ever existed," while Jefferson sided with John Adams in thinking that the British needed to purge their constitution of "its corruption, and give to its popular branch equality of representation." Hamilton dismissed this as "impractical," leading Jefferson to later claim in *The Anas* that "Hamilton was not only a monarchist, but for a monarchy bottomed on corruption."

Historians warn against making too much of the moment in history that pitted Jefferson against Hamilton, or of oversimplifying the character and the virtues of either man. In fact, I am not interested in making a case for the character of either man. Both men were brilliant, complex, and deeply flawed. But in the debate that has continued for two hundred years about the philosophical lineages of Hamilton and Jefferson, the latter has clearly emerged in the modern public consciousness as the winner. Today, we think of Jefferson as more urbane than Hamilton, more philosophical, more the gentleman—certainly not the kind of guy who would get lured into a duel with Aaron Burr. More than a million people each year visit the Jefferson Memorial in Washington, D.C. At the Treasury Department, which Hamilton founded, his statue isn't even prominently displayed in the front of the building. Instead, Thomas Jefferson's treasury secretary, Albert Gallatin (whom Hamilton once tried to have hanged), greets visitors on Pennsylvania Avenue. The historian Joseph Ellis has noted that there is more "talk" on the Internet about Jefferson than any other historical American figure. It's as if the more Hamiltonian we become, the more we yearn for the idealism of Thomas Jefferson. Yet what makes the Jefferson–Hamilton dichotomy so relevant and so troubling today is the way we so easily accept Hamilton as the *practical* man, the realist who knew manufacturing was inevitable in America, as were banking, corruption, and an ever-growing divide between the nation's wealthy few and its poor millions.

———

Were it not for Alexander Hamilton, I doubt Thomas Jefferson would ever have been president. After all, back in Virginia he had a beautiful hilltop home, two daughters, Martha and Polly, and Sally Hemings, the beautiful slave who gave birth to his other children. There was certainly no role that Jefferson liked better than that of the gentleman farmer. He would have finished out his days at Monticello had he not thought the nation was being pulled quickly away from pastoral republicanism. Nor was Jefferson alone in this thinking. The country had become divided between the Federalists, who believed in a strong central government, and Jefferson's Republicans, who believed men could be trusted to govern themselves.

Thus the campaign of 1800 grew vitriolic, and as Jefferson's most voluminous biographer, Dumas Malone, wrote, "It may be claimed that the personal attacks on the chief Republican were the most vicious in any presidential campaign on record." Though no one had yet found evidence of Jefferson's sexual relations with Sally Hemings, his allegiance to Thomas Paine had convinced most American clergy that Jefferson was an "atheist" and an "infidel." One minister wrote in the *Connecticut Courant*, "Do you believe in the strangest of all paradoxes—that a spendthrift, a libertine or an atheist is qualified to make your laws and govern you and your posterity?" Many preachers told their congregations that a vote for Jefferson was a vote against Christianity.

Jefferson, for his part, remained largely silent on the matter of his own beliefs, maintaining as he had all of his life that religion was a personal matter. But after he was elected president in 1800—after, as he wrote to a friend, the storm had passed and the ship of state had been "put on her republican tack"—Jefferson did make a private and personal response to those who questioned his religion.

Taking a pair of scissors to two King James Bibles, he created his own version of Christianity. Jefferson cut out all of the miracles—including the most important one, the Resurrection—then pasted together what was left and called it *The Philosophy of Jesus* (fifteen years later, in retirement at Monticello, he expanded the text, added French, Latin, and Greek translations, and called it *The Life and Morals of Jesus of Nazareth*). In an 1819 letter to William Short, Jefferson recollected that the cut-and-paste job "was the work of two or three nights only, at Washington, after getting through the evening task of reading the letters and papers of the day." Gone, for instance, is the first chapter of Luke, where the angel appears to Mary and announces the virgin birth and that her son "shall be called Son of the Highest." Jefferson began his private gospel with the more prosaic second chapter: "And it came to pass in those days, that there went out a decree from Caesar Augustus, that all the world should be taxed." Formally, Jefferson's gospel is a remarkably modern document in its almost cinematic scene-shifting and its collage-like composition. Jefferson begins the account of Jesus' baptism with two verses from Luke before cutting to this line from Mark—"John did baptize in the wilderness"— then shifting to Matthew's description of the Baptist's camel-skin raiments. Then, just as the Holy Spirit is about to descend like a dove, Jefferson cuts away again, and suddenly a very worldly Jesus is in Jerusalem, expunging the Temple of money-changers. If Jefferson didn't believe in the supernatural miracles, he did seem willing to believe that one man with a homemade whip could upend an entire Temple economy, then walk away unscathed.

Jefferson mentioned *The Philosophy of Jesus* in a few other private letters, but for the most part he kept the whole matter private, guessing (rightly) that the established church would simply see the compilation as another example of his "atheism." Nor did Jefferson later care to give Federalist newspapers another reason to remind him of his alleged sexual relations with his slave Sally Hemings, an entanglement certainly out of keeping with the philosophy of Jesus.

But Jefferson's severe redaction was probably a retaliatory act, as much as anything, against priests and ministers— "soothsayers and necromancers," Jefferson called them—who had unleashed the attacks on his character during the 1800 election. Jefferson believed that an authentic Christianity had long been hijacked by the Christian church. The teachings of its founder had become so distorted as to make "one half of the world fools, and the other half hypocrites." Jefferson would no doubt have agreed with Tolstoy that the Christian church had supplanted the Sermon on the Mount with the Nicene Creed to create a system of beliefs that Jesus himself wouldn't have recognized, much less laid claim to. "I abuse the priests indeed," Jefferson wrote to Charles Clay in 1815, "who have so much abused the pure and holy doctrines of their master." By stripping away the Gospel writers' claim that Jesus was the divine son of God, and by stripping away the subsequent miracles to prove it, Jefferson boasted that he had extracted "the diamonds from the dunghill" to reveal the true teaching of Jesus for what it was: "the most sublime and benevolent code of morals which has ever been offered to man."

Until then, Jefferson had claimed Epicurus as his patron philosopher. Two thousand years earlier, Epicurus had taught that life would be much easier to endure if we stopped fearing God and death—about which we can know nothing and do nothing—and followed instead a program of prudent self-sufficiency. "Everything easy to procure is natural, while everything difficult to obtain is superfluous," Epicurus wrote, adding that while natural wealth reaches a point of satiety, "the riches of idle fancies go on forever." Such a philosophy certainly would have appealed to Jefferson's agrarian vision for the new nation. But after the acrimonious 1800 campaign, Jefferson discovered that the philosophy of Epicurus didn't go far enough. "Epictetus and Epicurus give laws of governing ourselves," Jefferson wrote to William Short, "Jesus a supplement of the duties and charities we owe to others." Jefferson no doubt felt that not a few people owed him some charity.

Jefferson's tombstone at Monticello does not mention that the deceased was once president of the United States. Rather, it states that Jefferson authored the Statute of Virginia for Religious Freedom. In 1904, to honor the father of church–state separation, the Fifty-seventh Congress published *The Life and Morals of Jesus of Nazareth* for the first time, one hundred years after Jefferson pasted it together.

To read the Gospel story—the "good news"—through Jefferson's lens is instructive in a number of ways, the least of which is its representation of Jesus' "life." Most New Testament scholars agree by now that the infancy chapters of Matthew and Luke are pure myth. And no one has solved the mystery of the "missing years"—the two decades between when Jesus supposedly taught in the temple as a precocious child and when he came ambling along the Jordan River, asking to be baptized by the fiery zealot John the Baptist. From then until his execution a few years later, Jesus' life was a combination of walking, eating with followers and social outcasts, preaching, fishing a little, telling stories that no one seemed to understand, and offering largely unsolicited diatribes against the powers that be. That is to say, the life of Jesus, if unconventional, was nevertheless ordinary enough. Thousands of homeless men and women do pretty much the same thing every day in this country. But to find the *historic Jesus* within the fabulous accounts of the four Gospel writers is indeed an exercise of looking for diamonds in the compost heap.

Jefferson's gospel could not solve that problem. Nor did it need to. The life of this itinerant preacher was much less important to Jefferson than what he taught. *Somebody*, after all, spoke the Sermon on the Mount, or on the plain, or wherever it was spoken, and *somebody* told fascinating parables that explained little and left much up to the listeners—"those who have ears." What's more, Jefferson's objection to the version of Christianity taught in American churches was precisely that it did put so much more emphasis on Jesus' life and, consequently, his sacrificial death. By excising the Resurrection and Jesus' claims to divinity from his private gospel, Jefferson portrayed an ordinary man with an extraordinary, although improbable, message.

Indeed, reading Jefferson's gospel one hundred years after

its publication, it's hard not to become depressed, as did the "rich young ruler," about how nearly impossible Jesus' program would be to follow. To read the Gospels of Matthew or Luke is to be dazzled by one miracle after another. In that context, the actual teachings seem almost mundane ("Do that water-into-wine trick again!"). But to read Jefferson's version (what Beacon Press now publishes as *The Jefferson Bible*) is to face a relentless demand that we be much better people—inside and out—than most of us are. Which leads, as Jefferson must have suspected, to this unfortunate conclusion: The relevance of Christianity to most Americans—then and now—has far more to do with the promise of eternal salvation *from* this world than with any desire to practice the teachings of Jesus while we are here.

But Jefferson's gospel also leads to an impressive clarification of what those teachings are. One can make a list, and it need not be long. We should:

- be just; justice comes from virtue, which comes from the heart
- treat people the way we want them to treat us
- always work for peaceful resolutions, even to the point of returning violence with compassion
- consider valuable the things that have no material value

- not let the pursuit of wealth or the accumulation of possessions distract us from an authentic life, a true sense of being
- not judge others
- not bear grudges
- be modest and unpretentious
- give out of true generosity, not because we expect to be repaid

In all his teachings, the Jesus that Jefferson recovers has one overarching theme—*the world's values are all upside down in relation to the kingdom of God.* Material riches do not constitute real wealth; those whom we think of as the most powerful, the *first* in the nation-state, are actually the *last* in the kingdom of God; being true to one's self is more important than being loyal to one's family; the Sabbath is for men, men are not for the Sabbath; those who think they know the most are the most ignorant; the natural economy that birds and lilies follow is superior to the economy based on Caesar's coinage or bankers who charge interest.

Above all, this Jesus cannot abide hypocrites (from the Greek word for "actors"). He has nothing but contempt for men who would kill a woman because of adultery when they themselves have thought about cheating on their wives, or for temple officials who tithe mint and cumin but would do

nothing to help a poor woman with a child. "Stop *talking*
about righteousness," this Jesus is saying, "and *be* righteous."
It sounds simple enough. But of course it isn't simple at all.
Take the contemporary example of Martin Heidegger. No
twentieth-century philosopher, in my opinion, wrote with
more subtlety and richness on the protean subject of *being*.
But when it came time for Heidegger, as Germany's premier
thinker, to stand up to Hitler, he mutely accepted the Führer's
appointment to the rectorship at Freiburg University in 1933.
When heroism was needed most, Heidegger showed himself
to be a coward. Years later, concluding an essay on the uses of
poetry, Heidegger wrote, "The hard thing is to accomplish
existence." In this sentence, we might read a veiled apology
for his acquiescence under National Socialism. One can
talk about being, and one can talk about poetry, but to
accomplish being, to *act poetically*—that Heidegger could not
do. Nor, many have argued, could Jefferson, a slave owner
who as president did nothing to eliminate slavery. True
enough. But I would still argue that Jefferson offered a vision
for the country that was intrinsically religious because it was
intrinsically pastoral. And in that, Jefferson's vision was morally superior to the industrial capitalism that Alexander
Hamilton birthed upon the nation.

In his only book, *Notes on the State of Virginia*, published
the same year as Shays' Rebellion, Jefferson urged readers to

resist the factory life of large European cities and stay on the land. "Those who labor in the earth are the chosen people of God, if ever he had a chosen people, whose breasts he made his peculiar deposit of substantial and genuine virtue," Jefferson wrote in the chapter called "Manufacturers." Farmers intuit the laws of God within the laws of nature, and so become virtuous, he reasoned. They are, by the nature of their work, resourceful, neighborly, independent. They are the elemental caretakers of the world. Nor do they succumb to the crude opinions of the masses. "The mobs of great cities add just so much to the support of pure government, as sores do to the strength of the human body," Jefferson went on. But the farmer is freethinking and inquisitive. The manufacturer, by contrast, is a specialist, a cog, a wage slave. "Dependence," Jefferson concluded, "begets subservience and venality, suffocates the germ of virtue, and prepares fit tools for the designs of ambition." A manufacturer cannot be a citizen of a democracy, only a consumer within an oligarchy.

All his life, Jefferson trusted the laws of nature over the laws of the nation-state. Indeed, he wrote that trust into the first sentence of the Declaration of Independence, where it was "the laws of nature and of nature's God" that entitled the American colonies to break with Great Britain. Of Hamilton, his biographer Thomas Fleming observed, "As for his growing intimacy with the works of nature, the General was a disaster as a farmer." When the farm surrounding the Grange, his Manhattan estate, yielded only eighteen dollars one year, Hamilton remarked bitterly that he was as poorly fitted to farm as Jefferson was to be head of state. Four years after Jefferson made his pastoral plea in "Manufacturers," Alexander Hamilton submitted to Congress his *Report on Manufactures*, in which he dismissed Jefferson's agrarian vision in favor of developing industry, division of labor, protective tariffs, child labor, and prohibitions on many imported manufactured goods. That same year, to get small, generalist farmers and artisans off their land and into factories, Hamilton introduced the first tax on a domestic product—whiskey. Large, industrial distillers were charged a lower tax than the small-time producers. With their livelihoods on the verge of collapse, the smaller producers took up arms in what would become known as the Whiskey Rebellion. This played right into Hamilton's plan, which was to create a standing army that would defend federal laws as well as the bondholders who benefited from those laws that consolidated wealth in the hands of the few, large industrialists. Under Hamilton's instruction, the U.S. Army arrested suspected rebels and held them indefinitely without being charged. When Hamilton failed to produce legitimate suspects, he justified his actions to President Washington by claiming that the suppression of the rebels had scared any other potential troublemakers from standing up to the federal government. As the historian William Hogeland bluntly wrote, throughout Hamilton's career, he had one overarching goal—to establish "the essential relationship between the concentration of national wealth and the obstruction of democracy through military force." He achieved both, and in a spectacular fashion. For fiscal year 2008, the U.S. military budget was projected at around $480 billion, and the wealthiest one percent of Americans now earn more than the bottom 100 million Americans combined. Today, less than one percent of Americans work on farms, and many of those are huge, industrial farms responsible for an alarming amount of toxic runoff and erosion, stronger strains of disease, and climate change. That Jefferson's small-scale, self-reliant farmer is so unrecognizable to us today is evidence enough, should we need any, that we have inherited Hamilton's America, not Jefferson's.

In this context, the difference between Hamilton and Jefferson is the difference between a version of Christianity based on Jesus' life and death and resurrection, and one based on his teachings. Or to put it another way, it is a difference involving where one locates *basileia tou theou*—the kingdom of God. Is it, as Luke's gospel says, "in the midst of you," or is it, as John's gospel claimed, a reward saved for the sweet hereafter? To live by Jesus' *teachings* would be to live virtuously as stewards of the land; it would be to create an economy based on compassion, cooperation, and conservation; it would be to preserve the Creation as the kingdom of God. Jefferson was proposing a country of countrysides, a pastoral setting in which we would want to live. But to believe in Jesus' death as an event saving us *from* this world makes the abuse of it that much easier to justify. We are no longer stewards of the natural world, but exiles waiting for release. Hamilton was giving us a nation of factories from which we would want—perhaps in the end need—to be saved.

Hamilton's biographer Ron Chernow wrote recently, "Today, we are indisputably the heirs to Hamilton's America, and to repudiate his legacy is, in many ways, to repudiate the modern world." Yet in many ways, that *is* what we must do. Hamilton's modern world has led us to what one of America's most revered scientists, E. O. Wilson, calls the Moment—specifically, an unsustainable moment of overpopulation and overconsumption. It is a time in which serious decisions have to be made about the future of this country and how we conduct ourselves in that future. I am not suggesting that urban Americans should quit their jobs and become farmers; that could be equally disastrous. As Kentucky's most popular farmer, Wendell Berry, wrote in the short poem "Stay Home":

I will wait here in the fields
to see how well the rain

brings on the grass.
In the labor of the fields
longer than a man's life
I am at home. Don't come with me.
You stay home too.

But wherever we call home, *there* we must begin to confront the inescapable fact that Americans consume five times more resources than the earth can sustain. Our behavior will determine how many species survive or are lost, it will determine the rate at which global temperatures rise, and it will determine whether our own economy can come into line with the economy of nature. The poisonous discharge of Hamilton's industrial economy led us to this Moment.

Jefferson's alternative vision for the country—at once religious, pastoral, and truly democratic in scale—could lead us away from this perilous brink, into a future that finally translates his idealism into a sustainable, responsible reality.

II

Walt Whitman
at Furnace Mountain

Thomas Jefferson's redactions clarified for me the teachings that are at the heart of the Gospels—the difficult demands so often obscured in this country by mainline Christianity, which too often places its emphasis on matters (such as homosexuality and abortion) about which Jesus said absolutely nothing. But as Jefferson himself found, Jesus' teachings are essentially about how we should act toward others. There is very little instruction concerning how we might understand ourselves. Many social scientists have noted a fundamental difference between Eastern and Western ways of defining this mysterious fortress of molecules that we call the self. Westerners, we are told, define the self mainly through what Carl Jung called "extroversion," while Asian cultures put the focus on "introversion." Furthermore, said Jung, "for us [in the West], man is incommensurably small and the grace of God everything; but in the East, man is God and he redeems himself." Though this is certainly an oversimplification, my grandfather's fundamentalism had gone very far indeed in convincing me of my smallness, my sinfulness, my great need for divine intervention on my soul's behalf. When I turned thirty-three, the age at which my father killed himself, the entire set of pretenses that I called my *self* began to crumble. And I began thinking seriously about the Eastern alternative.

That is what sent me sailing down the Mountain Parkway of eastern Kentucky that same summer, in my old pickup, heading for a Buddhist monastery called Furnace Mountain. I followed the interstate for twenty miles past rolling cattle operations and black tobacco barns. I pulled off at the Clay City exit and drove through a narrow little community wedged between the two sets of hills. I had made arrangements through a friend to spend the summer in a small cabin that sat neglected in a lower corner of the Furnace Mountain property.

The monastery itself clings to the side of a ridge near Kentucky's Red River Gorge, and it overlooks Clay City—a town known more for its drag strip than for its dedication to Buddhism. The monastery was named for a nineteenth-century iron smelter that still stands near the summit. I call Furnace Mountain a monastery (which at one time it was poised to become), but really it is a retreat center where anyone is welcome to participate in the weekend *zendo*, a meditation service, or to actually live there and study with Dae Gak, the Zen master who started Furnace Mountain in the early 1980s.

I first learned of it when my friend Chris brought me hiking there in the spring. We parked my truck near the road and started up a narrow trail that wove around huge slabs of sandstone and toward a ledge dominated by mountain laurel and cedars. From that perch, we could look down at the entire county. Hawks circled beneath us. We could hear nothing but the wind as it pulled apart some prayer flags that visiting Tibetan monks had strung up a few months before. We descended by another route, and I was pushing through a dense thicket of young trees when I saw a deep-hued patch of blue shining below. To reach level ground, we had to lower ourselves into a crevice between two huge sandstone boulders and carefully shimmy down. Then, when we stepped out into the clearing, I saw it. Squeezed onto a small plateau right beneath a steep abutment, this blue-roofed Buddhist temple looked about as unlikely a sight as one could imagine in rural Kentucky. The corners of the ceramic tile roof flared out into soft curves. The temple was framed by the thick trunks of Georgia pines, stripped of their bark and stained. A parapet made of flat stones wrapped around the foundation of the temple.

We climbed a long set of steps dividing a slope covered with pink vetch. Inside, the walls were white. A scroll painted with the Zen master's calligraphy covered the electrical panel. Against the back wall stood an altar, on top of which a golden Buddha rested on a lotus leaf, his legs folded and his hands set in the teaching position. (Some local drunks once tried to carry the Buddha off until Dae Gak convinced them that if it was real gold, they wouldn't be able to lift it anyway.) The Buddha was surrounded by symbols of the four elements: a bowl of rice (earth), an incense burner (wind), two candles (fire), and a bowl of water. Meditation cushions formed a neat U around the altar. Here, master and students sat for hours, as still as possible, emptying their minds. I sat down on one, crossed my legs, and instantly felt like an impostor.

We quickly left and headed back to my truck. That was when Chris told me about an abandoned cabin somewhere at the bottom of the property. We passed the "teahouse," a rectangular building with an open eating area in the middle and bedrooms at each corner. Here Dae Gak lived with his wife, Mara, and their fourteen-year-old son, Sam, a student in Chris's Montessori class. But at that moment, Sam was standing on the porch of the teahouse, firing Idaho potatoes from a grenade launcher he had fashioned out of PVC pipe and duct tape, powered by ignited hair spray. Across the gravel driveway, a stand of trees heroically held their ground as potatoes splattered against their trunks. Chris and I examined Sam's artillery until our voices were drowned out by the sound of a twin-engine motorcycle barreling toward us. The rider pulled out beside us, shook off his helmet, and grinned. It was Dae Gak. He had closely cropped gray hair, and wore jeans and a blue flannel shirt. In a lineup, he would be the last person I'd pick as the Zen master.

He suggested Chris and I stay for dinner, and as we sat around the teahouse with Dae Gak's family and his students, I asked him about the difference between Zen and other kinds of Buddhism. "Zen is about not-seeking," Dae Gak shot back. It struck me then as an odd thing to say. What was everybody doing here, if not seeking? Didn't they all want something they hadn't found in their own Protestant, Catholic, or Jewish upbringings? Didn't they want what the Buddha promised—enlightenment? Or were they seeking to no longer seek? I wasn't sure, but the idea fascinated me. To *not* seek, to simply accept the world, in Ludwig Wittgenstein's phrase, *as we find it*—was it really that simple?

Chris asked Dae Gak about the abandoned cabin. He said an Austrian man had built it eight years earlier, but fell into financial hard luck and ill health before he could add electricity or water, if indeed that had ever been his plan. He moved back to Austria and had never spent a night in the cabin. It had been sitting empty for a couple of years.

As we were leaving, I got up the nerve to ask Dae Gak if I might stay in the cabin that summer. I didn't mention that I was in the middle of, and doing my best to mask, some kind of psychosomatic breakdown, that my skin crawled and anxiety attacks constantly leveled me. I didn't say that I was most likely suffering from survival guilt, having outlived the age of my father when he shot himself. None of that seemed appropriate. Rather, I said that I wanted to think and write about Eastern influences on the American poet Walt Whitman, and that an American Buddhist monastery seemed like the right place for such musings, which was true. Dae Gak considered this, then fetched me the key to the cabin.

I was immediately drawn to Whitman when I started reading him in college, because he represented such a clear antidote to the repressive Christian morality that I blamed for my father's suicide. Sin lurked everywhere, according to my grandfather's theology, and guilt was the surest way to guard against it. When my father was in seminary, he wouldn't go to movies with his friends, because to my grandfather Hollywood represented the worst of moral decrepitude. My father wanted to go, my mother told me, but he was afraid my grandfather would ask if he had gone, and he couldn't bear the thought of having to admit such a transgression.

As for me, throughout my late adolescence, I gradually began to suspect that all of this moral heaviness, this dreary goodness, was not the founder's invention, but rather a notion thought up by Jesus' zealous proselyte the apostle Paul. It was Saul of Tarsus who began to wonder: If Jesus had replaced the Mosaic Law with the Holy Spirit, what would stop this new church from veering into anarchic hedonism? Indeed, he was already hearing rumors to that effect coming from the early churches in Corinth and Rome. Moreover, the Gentiles in Rome had no Law. What would save them from the wrath of God? Paul worked out a rather tortured theology whereby they would be saved by the blood sacrifice of Jesus, now called the Christ. If all humankind had inherited the sin of one man, Adam, then all could be cleansed of it by the "ransom" of one man, Jesus. There is no place in the Gospels where Jesus calls himself the Christ—indeed he explicitly rejects the title several times—nor is there any place where he says that his death will forgive sin. But Paul became obsessed with this Judaic idea of the sacrificial lamb, the servile martyr. To have "faith" in this belief of a blood sacrifice would replace a need for the Law. Believers would receive instead the gift of grace. But there was still a problem here. If we Gentiles have grace, why not keep sinning? It can, after all, be a lot of fun (think of Bessie Smith singing, "Slay me, 'cause I'm in my sin / Slay me 'cause I'm full of gin"). So Paul wrote in his letter to the Roman church:

> What shall we say then? Are we to continue in sin that grace may abound? By no means! How can we who died to sin still live in it? Do you not know that all of us here who have been baptized into Christ Jesus were baptized into his death? We were buried therefore with him by baptism into death, so that as Christ was raised from the dead by the glory of the Father, we too might walk in newness of life.
>
> (Romans 6:1–4, RSV)

Just as Jesus rose from the dead, so we rise out of sin into Spirit. Finally, for Paul, this law of the Spirit replaced the Law of Moses. "Any one who does not have the Spirit of Christ does not belong to him," he wrote. "But if Christ is in you, although your bodies are dead because of sin, your spirits are alive because of righteousness" (Romans 8:9–10, RSV). In his letters, Paul grew more and more obsessed with this schism between sin and the law of the Spirit. While Jesus mentioned the word "sin" only a handful of times in the Gospels, Paul repeated it at least ninety times throughout his letters. And he never forgot that the first sin was a sin of the flesh—nakedness. "If you live according to the flesh you will die," he warned the Roman churches (Romans 8:13, RSV). While Whitman wrote in his long poem *Song of Myself,*

> I am the poet of the body,
> And I am the poet of the soul,

Paul drew an utter distinction between the two realms. Just as Whitman became the exemplar of what William James called *the religion of healthy-mindedness,* Paul introduced into Christianity the first seeds of the *sick soul.* "For I know that nothing good dwells in me," he wrote (Romans 7:18, RSV). Whereas Whitman sang of an oceanic union with God, a feeling of infinite largeness, Paul remained fixated on an awful separation.

In the end, Paul managed to take the "good news" of the Gospels and turn it into another stifling set of prohibitions, laws utterly lacking in spirit. But how would he enforce them? Paul didn't have a Roman army to keep his followers in line. Instead he erected what amounted to an internal police state that was constantly on guard against sin—guilt. Christians could police themselves! This was far better than any army.

Guilt provides around-the-clock surveillance; even in dreams one isn't free. As anyone who has grown up in a Baptist church knows (and I suspect it is not particular to that denomination), the atmosphere of guilt can become so pervasive that one doesn't even have to *do* anything wrong. We were *born*, after all, with the mark of sin. Just being alive, just waking up in the morning, becomes a dubious endeavor, a transgression of the flesh. Such at least became my own experience, and I think my father felt this much more profoundly. This was the Christianity that Paul had given us.

One day in the mid-1980s, while I was attending a college journalism convention in Washington, D.C., I wandered into a used-book store and picked up a paperback called *Zen Buddhism* by D. T. Suzuki. I opened it and read the first, startling sentence: "Zen in its essence is the art of seeing into the nature of one's own being, and it points the way from bondage to freedom." I knew nothing about Zen or Buddhism, but I was ready for some freedom from what the poet William Blake called the "mind-forg'd manacles" of Pauline guilt. I bought the book and went back to my hotel room to read. In a later chapter, Suzuki quotes the Sixth Patriarch of Buddhism, Huineng: "Think not of good, think not of evil, but see what at this moment thy own original face doth look like, which thou hadst even prior to thy own birth." This image of some original face, some original self free from the laws of good and evil, startled me, and though I could not articulate it to my-self, that image was my first clue to what Suzuki meant about Zen pointing a way from bondage to freedom. We do not have to divide the world up into good and evil, black and white, heaven and hell, moral and immoral, body and soul, male and female, straight and gay. We have only to truly see the world as it is—neither good nor bad, but simply *given*. To see things this way, I realized, was no longer to refract everything through the lens of Christian judgment. Sin and temptation were not constantly lurking in this fallen realm, as my grandfather be-lieved. Zen seemed to be replacing *original sin* with something called *original nature*. The point was to rediscover—through sitting, chanting, bowing—this face we had before we were born. I had come to Furnace Mountain to see if this way of thinking, this way of being in the world, might set me on a new tack. And I was proceeding on the hunch that Walt Whit-man could help—he had, after all, written the great American poem on the subject, *Song of Myself*. Perhaps that was the new gospel I needed, a particularly American gospel. Alone at Furnace Mountain, I could embark on a thorough, even mo-nastic, study of that text.

So that May, I hauled my truckload of supplies—a duffel bag packed with clothes, a box of books and notebooks, and lots of rice—into the sparsely furnished cabin. Some chinking was falling away from between hand-hewn cedar logs. The front door was a work of art: a tightly fitted puzzle of dovetailed oak boards. Inside, a woodstove split a main room from a kitchen. With no electricity, water, or toilet, the hut made me feel like a mountain recluse fleeing some toppled dynasty. A makeshift set of stairs (more of a ladder, really) led to the loft, where mice had eaten small chunks of foam out of the mattress of a single bed. I unpacked my gear and set up a small shelf of books on the desk that sat in front of the cabin's main window. This was my personal syllabus for the summer:

- *Leaves of Grass*, Walt Whitman
- *The Writings of Ralph Waldo Emerson*
- *The Varieties of Religious Experience*, William James
- *Going Beyond Buddha*, Robert Genthner and Dae Gak
- *One Hundred Poems from the Chinese*, Kenneth Rexroth
- *Zen and the Birds of Appetite*, Thomas Merton
- *Microcosmos*, Lynn Margulis and Dorion Sagan

In some way I hadn't quite articulated to myself, I felt that these books—poetry, philosophy, theology, and science—were all speaking to the same problems, which I dimly understood to be my own.

I walked out into the clearing in front of the cabin and called Dae Gak to let him know I had arrived. "I'm on a cell phone my friends made me buy," I explained by way of apology for what seemed to me a rather un-Zen-like device.

"Oh, that's good," he said. "I'm glad you got one. If you get bitten by a copperhead, just call us and we'll come get you."

That summer, I rousted myself awake at five-thirty each morning and climbed the serpentine path leading up to the temple. With Dae Gak and a few of his students, I would sit on a cushion for two hours, counting my breaths. I had read that such meditation was a good cure, or at least a palliative, for anxiety and depression. And I thought it might help me understand the more Eastern side of Whitman's poetry. To simply *be*, fully present in one place and time: that is the point of this meditation. But I found the process incredibly difficult. Here I was, sitting cross-legged in a beautiful temple, surrounded by mist-covered mountains, but I couldn't seem to quiet my mind. I concentrated on my breathing for a few minutes. But then something would distract me—something I had left undone back in town, some song lyric, or even the idea of instant cheese grits for breakfast. I couldn't stay in the moment. I couldn't sit still.

As a people, we Americans are not a contemplative bunch. We can handle a Wednesday-night church supper or a short

Sunday sermon, but beyond that we grow restless. We respond best, as Jung had noted, to external stimuli and external rewards, and we have built an entire economy around that psychology. Most of us simply are not very good at sitting still. Aside from fringe Protestant movements like the Quakers, we do not listen to the silent voice that might emanate from within if we paused to listen. Restlessness. That's what Alexis de Tocqueville said our problem was. In *Democracy in America*, he wondered why Americans were so restless in the midst of their prosperity. "In the United States a man builds a house in which to spend his old age, and he sells it before the roof is on," Tocqueville observed. Some sociologists have reasoned that since the United States was founded by immigrants, it makes sense that we are always on the move. Others have suggested that the size of the country leads to a certain frontier mentality—the urge to "light out for the territories," as Huck Finn planned to do at the end of his own story. We do not, like many Europeans, sense a long communal history that holds us to one place, one city, one town. So we gave the world the automobile as a testament to our restlessness. Early on, Tocqueville noticed that Americans' geographical restlessness was reflected in our "taste for physical gratifications": "He who has set his heart exclusively upon the pursuit of worldly welfare is always in a hurry, for he has but a limited time at his disposal to reach, to grasp, and to enjoy it." I had come to Furnace Mountain from that culture, thinking I could escape its impurities, its superficial gratifications, its frenetic pace, only to find that after just a couple of weeks, *I wanted all of those things back*. Or at least some of them. I had fooled myself into thinking I could shed so quickly my American skin.

Gradually, though, I felt myself slowly becoming more focused during the morning sittings. I was able to remain still more easily and to let whatever thoughts come and go without getting caught in a mesh of distraction. I felt I was beginning to understand more intuitively what this practice was about. To breathe is to participate in the fundamental activity of the natural world—the transfer of carbon into oxygen and back. "Inspiration and respiration," as Whitman writes in *Song of Myself*. That breath was the invisible syntax that held it all together. Feeling proud of my discovery, I asked Dae Gak after one morning session if I was doing the right thing, concentrating on this Universal Breath.

"That's a thought," he said. And a thought is the heresy of Zen. "Of course you are participating in that," he went on, "but you're not thinking about it."

I asked him what the perfect state of mind would be.

"Just presence," he replied.

After the morning sittings, I walked back down to the cabin and made coffee and grits on my small propane stove. I sat on the front porch with my journal and *Leaves of Grass* and waited for the sun to appear behind the eastern ridge. By then the valley was full of birdsong. Indigo buntings darted over the high grass down below the cabin's porch. I had nowhere I needed to be, and only a few friends actually knew where I was. During the chilly May mornings, I pondered the rambling introduction to the first edition of *Leaves of Grass.* There, Whitman sets out to define the task of "the great poet," presumably so that his readers will know him when they see him in the pages that follow. That is to say, Whitman was teaching American readers how to read a new kind of poem, how to understand a new kind of poet. He, the great poet of the long poem, would not be eulogizing mythic heroes or conjuring dramas of heaven and hell. Rather, he meant to celebrate the "demonstrable," "the essences of the real things," "the common people," "the roughs and beards and space and ruggedness and nonchalance that the soul loves." He was going to write, above all, a *folk poem* of great equanimity. And because "the fruition of democracy lies altogether in the future," he would offer a portrait of America not as it was but as it might be. In his later essay "Democratic Vistas," Whitman wrote that Americans must learn to travel "by maps yet unmade." He aimed for *Leaves of Grass* to be the first great map of a truly democratic landscape.

By 1855, when the first edition was published, Whitman had utterly lost faith in party politics and the office of the presidency, and he didn't believe the American version of democracy was even worthy of the name. He had observed the astonishingly ineffectual presidencies of Millard Fillmore, Franklin Pierce, and James Buchanan. After the passing of the Kansas–Nebraska Act in 1854, which opened the western territories to slavery, Whitman offered this unrestrained assessment of Pierce: "The President eats dirt and excrement for his daily meals, likes it, and tries to force it on The States." That same year, the runaway slave Anthony Burns was sent back south by a Boston judge in accordance with the Fugitive Slave Act. The return of Burns to his slave owner enraged Whitman, as it did his friend Henry Thoreau, who called a town meeting to inform his neighbors that they were cowards and hypocrites who might benefit from reading their Bibles once in a while. What's more, the economy was crashing in 1854, and the gap between the rich and the poor was wider than it had ever been. The wealthiest one percent of Americans owned thirty percent of all wealth. Whitman blamed this miscarriage of idealism, as had Thomas Jefferson, on the "vast ganglions of bankers and merchant princes." He lamented that America had produced no great literature, but only foppish imitations of European writing. What passed for culture in the United States was simply pretentiousness. There was no question in Whitman's mind: America needed him and needed him now. (At Furnace Mountain, as I settled into a serious consideration of *Song of Myself,* I started to agree that, for all of Whitman's churlish self-promotion, he was right. We did, and we still do, need him as the national poet best equipped to show us our better selves.)

In the introduction to the first edition of *Leaves of Grass,* Whitman claimed that he would dismiss priests and politicians and replace them with a new hero, the *common man.* Whitman would praise the "wellhung man" over the priest because "every man shall be his own priest." Certainly no project could be more democratic than that. Ralph Waldo Emerson had first introduced this radical idea in an 1838 commencement address that mightily upset his audience, the graduating class of Harvard Divinity School. In that speech, Emerson advised the young ministers to renounce preaching the "tropes" of the Gospels and instead point their parishioners back toward their own "divine nature." The problem with the established church, Emerson charged, is that it teaches our smallness instead of our largeness. "In how many churches," he asked, "by how many prophets, tell me, is man made sensible that he is an infinite Soul; that the earth and heavens are passing into his mind; that he is drinking forever the soul of God?" For Emerson, this infinite Soul, or Oversoul, could always be found reflected in the individual minds of all men and women. With breathtaking sweep, he was pushing aside the legacy of American Puritanism and replacing it with a new religion, Transcendentalism. He was replacing the church's emphasis on *sin* with the individual's concern for his or her own *soul.* Jesus, he said, was ravished by the soul's beauty—"he lived in it, and had his being there." He had climbed to the fountainhead, the fundamental intuition. "One man was true to what is in you and me," Emerson concluded. That one man, a Mediterranean wanderer, saw that "God incarnates himself in man, and evermore goes forth to take possession of the world." Emerson did not, like Thomas Jefferson, deny Jesus' divinity; he simply said the same divine potential resides in every human heart.

That idea set the foundation for *Song of Myself.* Perhaps the only other text, besides the Divinity School address, that had as much influence on Whitman was Emerson's lecture "The Poet." Here we find Emerson at his most vatic, presenting "the Namer" as the priest's replacement, the man or woman who "will reconcile me to life and renovate nature." Emerson's transcendentalism was never a doctrine of *transcending* the manifest world, but rather of infusing every inch of it *with* transcendence, with what Emerson called "a divine aura." I think he would have agreed with the Buddha that life is suffering—*dukkha*—because we no longer listen to our soul, or alternately, our Buddha nature. Emerson cast this same sentiment in the language that his American audience would understand: "For it is dislocation and detachment from the life of God that makes things ugly," and therefore it is the job of the poet to "re-attach things to nature and the Whole." This Whitman seemed to take as his charge, just as he must have certainly aspired to Emerson's claim—the one he had been working up to throughout the essay—that "poets are liberating gods." Why are poets liberating gods? Because, said Emerson, "they are free, and they make us free." How do they make us free? "The use of symbols has a certain power of emancipation and exhilaration for all men." The true poet shows us the world anew through the symbol, or the poetic image. And that symbol becomes a threshold through which we can again see the world, and return to it, but return more intensely.

After twenty pages of celebrating the form and function of this higher man, Emerson stopped to admit, "I look in vain for the poet whom I describe. We do not with sufficient plainness or sufficient profoundness address ourselves to life, nor do we chant our own times and social circumstance. If we filled the day with bravery, we should not shrink from celebrating it. Time and nature yield us many gifts, but not yet the timely man, the new religion, the reconciler, whom all things await. . . . Yet America is a poem in our eyes; its ample geography dazzles the imagination, and it will not wait long for meters." Emerson had only to wait a decade before a volume of poetry—plain and profound—arrived in the mail. That first edition of *Leaves of Grass*, made up mostly of *Song of Myself*, annihilated the derivative Anglophile verse of Whittier and Longfellow, and Emerson hailed it for what it was: "the most extraordinary piece of wit and wisdom that an American has yet contributed." (It remains both as extraordinary and as relevant today. And it remains the most original and inspiring contribution to the apocryphal canon that I call the American Gospel.)

There is little evidence in Whitman's biography before 1855 that he would become the poet behind *Leaves of Grass*. In the 1820s, his father, a carpenter, moved the family to Brooklyn to take advantage of the building boom there. But Walter Whitman did not prosper though his family swelled to nine children, one of whom was mentally retarded. His son Walt was a mediocre student. At age eleven, Walt apprenticed as a printer to help support the family, and later went to work as a journalist for various newspapers and as a schoolteacher. He even wrote a temperance novel that years later he tried to disavow. But between the writing of conventional fiction and poetry in the 1840s, and the publication of the first edition of *Leaves of Grass* in 1855, something profound and monumental happened to Walt Whitman. No one really knows what, and it ultimately doesn't matter. What does matter is that by 1855, Walt Whitman had reinvented himself as a truly original American genius. When Sojourner Truth heard *Leaves of Grass* read aloud, she said of Whitman, "Never mind the man's name—it was God who wrote it."

We do know that in 1854, Whitman was fired from the *Brooklyn Daily Eagle* for opposing the expansion of slavery into the western territories. By then, he had also learned carpentry and had built and sold a few small houses in Brooklyn. But according to his brother George, Walt was a bit too much of the contemplative loafer to ever make a serious go of it as a builder. Still, the working-class experience seemed to transform Whitman the journalist, who had at times been condescending toward the lower classes, into the poet of the common man. And by the third line of *Song of Myself*, he is already proclaiming, "For every atom belongs to me as good belongs to you." On a purely scientific level, this would turn out to be true—the famous flapping of a butterfly's wing that affects the farthest star. The discovery that "the active state of the atoms of one body has an influence upon the atoms of a body in contact with it," Whitman took from the German chemist Justus Liebig, whose work became popular in America in the 1840s. But Whitman quickly converted that science into his expansive definition of the self. Because the first edition of *Leaves of Grass* was published anonymously, we don't learn the identity of its author until his grand entrance into section 24: "Walt Whitman, an American, one of the roughs, a kosmos . . ." Suddenly, the poet is at once a common man and a divine emanation of the universe. At first, the interdependent atoms stand as emblems of radical democracy: "By God! I will accept nothing which all cannot have their counterpart on the same terms." And then the section takes a raucous turn as Whitman begins to expound on the cosmic nature of his fleshy self:

> Divine am I inside and out, and I make holy whatever
>   I touch or am touched from;
> The scent of these arm-pits is aroma finer than prayer,
> This head is more than churches or bibles or creeds.
> If I worship any particular thing it shall be some of the
>   spread of my body. . . .

As he begins to enumerate the parts of his body worthy of such worship, those lineaments begin to blend beautifully into the earth itself:

Trickling sap of maple, fibre of manly wheat, it shall
  be you. . . .
You sweaty brooks and dews it shall be you,
Broad muscular fields, branches of live oak, loving
  lounger in my winding paths, it shall be you. . . .

Whitman shared Emerson's view that the natural world is a
manifestation of the Over-soul. Whitman used the term *kosmos* to emphasize the Greek sense of "unity," the "harmony"
of that world. And if the soul is good, so is the *kosmos*, and
by extension, the flesh. "Not an inch nor a particle of an inch
is vile," he maintains at the beginning of *Song of Myself*.
Later, he announces his intention to reveal to all Americans
their sacred nature:

I am the mate and companion of people, all just as
  immortal and fathomless as myself;
They do not know how immortal, but I do.

The great poet of the people would show them the divine
nature that lies hidden within everyone. That had been Emerson's advice to the graduating Harvard ministers in 1828,
but we can be relatively certain that only Whitman took seriously such a heretical charge. Yet set within the context of
Buddhism, it no longer seems quite so scandalous. It seemed,

in fact, to be the impulse that drew Dae Gak's students, and also drew me, to Furnace Mountain. It was, of course, exactly the opposite injunction that my grandfather had given in his sermon "Open Eyes," where "the flesh" could never be trusted, where "man is merely clay." But that theology had not served my father in the end, and when I was thirty-three, it wasn't helping me either. I loved my grandfather immensely, but at Furnace Mountain, I was searching for some antidote to his oppressive faith.

For Whitman to understand his own divinity and that of others seems very much like students of Zen discovering that they possess Buddha nature. To understand this is to see the world as Whitman does, as a subject worthy of praise. He calls himself "the caresser of life wherever moving," and *Song of Myself* is a celebration of a physical landscape—a nation, not a state—and its people, from prostitute to slave to a president who passes by in his carriage and (imagine this) lifts his hat to the poet watching from the street. And as the caresser of life wherever moving, Whitman also sets himself up as a reconciler at every level—from the atom, to the individual, to the nation, to the *kosmos*.

There's a story I've come across several times in the literature of Zen Buddhism. A young monk approaches his master and says that he is ready to receive the master's teaching. "Have you eaten your breakfast?" the master asks. The young monk says that he has. "Then wash your bowl," the teacher replies. That's it—wash your bowl, let the everyday be its own kind of enlightenment. I tried to conduct my days at Furnace Mountain in the spirit of that idea. Each morning after breakfast, I retrieved two gallons of water from the stream, then washed my dishes and filled the solar shower. I swept the wooden floor and cleaned up after the mice. And I tried to do it all with the same deliberateness and attentiveness that sent Thoreau out to Walden Pond and that brought Dae Gak's students to Furnace Mountain.

After the household chores were done, I pulled on my hiking boots and went rambling over the six hundred–plus acres of sandstone bluffs and dramatic cliffs that make up Furnace Mountain.

Not many lines of American poetry get quoted more often than these from the beginning of *Song of Myself*:

I loaf and invite my soul,
I lean and loaf at my ease . . . observing a spear of
summer grass.

Whitman shrugs off the negative connotations of "loafing" by turning it into a spiritual exercise. Here is a man who had abandoned days measured by clocks and watches to find his soul's voice written in the "uniform hieroglyphic" of summer grass. It was easy to feel a similar freedom and exhilaration in the woods around Furnace Mountain. In a letter to his brother, Henry, William James wrote that the hikes he took in the Adirondacks were his "main hold on primeval sanity and health of soul." I felt that keenly as I scrambled over the boulders and climbed the ridge sides of Furnace Mountain (only rarely did I have to resort to my portable pharmacy to battle an anxiety attack).

Sometimes, when I reached the top of a capstone, I would find a flat spot, cross my legs, and count my labored breaths. I imagined myself to be one of those ancient Chinese poets, wanderers who had abandoned their bureaucratic jobs as court scriveners to build small huts in the mountains and dedicate their lives to writing short poems about waterfalls and shifting fog. One thing that is remarkable about the deep gorges and dense forests surrounding Furnace Mountain is the resemblance they bear to the landscape of southern China, where Zen Buddhism took hold in the first century CE and where those poets of the Tang Dynasty wrote the greatest nature poetry we have. Two-thirds of all the wild orchids in central Appalachia, where Furnace Mountain sits, are cousins to those in China. There are only two species of tulip poplar in the world—one in China and one in the eastern United States. In both forests, the nonwoody plants have developed underground storage systems, and most of them bloom in early spring, before the canopy closes over them. Inspect the forest floor in southern China and Appalachia, and you will find the same mayapple and jack-in-the-pulpit, the same ferns and ginseng. Apparently, what connects these two ecosystems on opposite sides of the globe is that neither suffered extensive glaciation during the Pleistocene era. When the ice withdrew, only these two regions retained the plant diversity that was once characteristic of each entire continent. And what connects the great Tang poets to Whitman is not only that they were inspired by such similar landscapes, but also that they shared an antipathy for speaking about the metaphysical realm: about things one cannot know, one should remain silent. "All truths wait in all things," Whitman wrote in the middle of *Song of Myself*,

> Logic and sermons never convince,
> The damp of the night drives deeper into my soul.

This is, above all, a poetry of belonging, a poetry of elemental contact with something much larger than the self. Because, says Zen Buddhism, true belonging can exist only prior to the cleavage of subject from object, the poets wrote out of an intense identification with their native landscapes. Thus "The

American Scholar," Emerson's famous Phi Beta Kappa address at Harvard, began by urging that "the ancient precept, 'Know thyself,' and the modern precept, 'Study nature,' become at last one maxim."

Whitman, in his own wildly loquacious way, with his long lines and elliptical sentences, was creating a poem and a philosophy that resembled this Eastern attitude much more than it did his Western predecessors'. He was not, like Milton, trying to justify the ways of God to man. Though Whitman said he saw the *proof* of God in every blade of grass, he was as adamant as the Chinese masters that he knew nothing *of* God. And like the ancient poets, Whitman spends whole sections of *Song of Myself* merely naming—listing—species, classes of people, mountains, and rivers. "My palms cover continents," he writes in the poem's longest section. And then, like the ecstatic saints who were rumored to physically levitate while contemplating God, Whitman himself takes to the skies for an aerial view of his beloved nation:

> Over the growing sugar . . . over the cotton plant . . .
>   over the rice in its low moist field;
> Over the sharp-peaked farmhouse with its scalloped
>   scum and slender shoots from the gutters;

> Over the western persimmon . . . over the longleaved
> corn and the delicate blue-flowered flax;
> Over the white and brown buckwheat . . .

and so on for pages. Whitman is sometimes faulted by critics for this unrestrained penchant to enumerate. It is as if he took seriously the Buddhist metaphor that the world is made up of "ten thousand things" and he set out to name them all, or at least all of the American things. It also reflects the naturalist's impulse to collect and categorize. But Whitman's classifications do not resemble the dry taxonomy of science. In "The American Scholar," Emerson asked rhetorically: "What is classification but the perceiving that these objects are not chaotic, and are not foreign, but have a law which is also a law of the human mind?" *Song of Myself* performs such a classification. One senses an entire ecology at work within the poem. The song itself holds together the poem's ten thousand things so that they function, like an ecosystem, as one vast interdependent whole. Furthermore, to collect such a variety of places, people, flora, and fauna within *Song of Myself* signified that the most ordinary of objects was worthy of praise: "There is no object so soft but it makes a hub for the wheeled universe," wrote Whitman. With that sentiment, he was foreshadowing the work of the great American naturalist Aldo Leopold, who wrote, "The outstanding scientific discovery of the twentieth century is not television, or radio, but rather the complexity of the land organism. . . . If the land mechanism as a whole is good, then every part is good, whether we understand it or not."

Whitman keeps insisting throughout *Leaves of Grass* that the book the reader is holding is *not* a poem but only a directive pointing to the *real poem* at the edge of the page—what his friend Thoreau called "the poem of creation." In "Song of the Rolling Earth," Whitman seems on the verge of abandoning language altogether:

I swear I begin to see little or nothing in audible
  words,
All merges toward the presentation of the unspoken
  meanings of the earth,
Toward him who sings the songs of the body and of
  the truths of the earth,
Toward him who makes the dictionaries of words that
  print cannot touch.

The poem, if it is to be of some use, has to reconnect the bodily reader with the physical world. Thus Whitman refused to speak in a conceptual language untethered from experience. In "Starting from Paumanok," he asks the reader:

Was somebody asking to see the soul?
See, your own shape and countenance, persons,
    substances, beasts, the trees, the running rivers,
    the rocks and sands.

It's a response you might expect from a Zen master. Within Zen, the snake cannot be seen as evil; it must simply be accepted for what it is—something one best not step on. In the West, thanks to the trickster serpent of Genesis, we claim to have eaten from a Tree of Knowledge, and that knowledge is speculative, analytical, conceptual. Once the First Parents ate of that fruit, they ceased to be poets—true namers, creating words out of breath and the instruments of their own bodies. They no longer *sang* the world. And when their praise song ceased, the poem of Creation slowly fossilized into a collection of mere matter.

That Western perspective proved to be long-lived. By the seventeenth century, John Locke wrote, "The intrinsic natural worth of anything consists in its fitness to supply the necessities or serve the conveniences of human life." The industrialism that followed championed scientific and economic values at the expense of all others. Except for some Romantic poets, few questioned the "dark Satanic mills" that were beginning to crowd out European pastures. But when the natural world is drained of its aesthetic and poetic values, the mountains and rivers that inspired the Chinese masters are reduced to one thing, a "natural resource." A particular landscape becomes an economic abstraction. And what follows is never pretty.

Fifty miles east of Furnace Mountain, coal operators are obliterating the central Appalachian mountain range in order to extract coal as quickly and as cheaply as is technologically possible. The modern industrial economy is turning the mixed mesophytic forests of southern Appalachia into a landscape whose value can only be measured in tons of coal or millions of dollars. That is to say, these landscapes can only be measured by the abstract language of the economist, a language that is the opposite of poetry.

Throughout *Song of Myself,* Whitman expresses an impatience with Western man's preoccupation with those two unshakable questions: "Where did we come from?" and "What happens when we die?" He writes,

I have heard what the talkers were talking . . . the talk
   of the beginning and the end,
But I do not talk of the beginning or the end.

There was never any more inception than there is now,
Nor any more youth or age than there is now;

And will never be any more perfection than there is now,
Nor any more heaven or hell than there is now.

"Just presence," as Dae Gak had said. To be present in the present—fully, attentively, amorously—that was Whitman's way as well. And his friend Thoreau wrote that to live in such a way was to live by "a newer testament—the gospel according to the moment." Such a new gospel replaces the linear, Christian notion of history, which began with Adam and Eve and ends on the Day of Judgment, with a cyclical understanding, measured instead by changing seasons and cycles of life. Whitman calls into doubt the eschatological belief in the Last Days, an apocalyptic end followed by eternal damnation or salvation. "Why should I wish to see God better than this day?" he asks.

I see something of God each hour of the twenty-four,
    and each moment then,
In the faces of men and women I see God, and in my
    own face in the glass;
I find letters from God dropped in the street, and
    every one is signed by God's name,
And I leave them where they are, for I know that
    others will punctually come forever and ever.

By extracting God from the afterlife and inserting him into every atom of this world, Whitman in effect cancels the need for a Heavenly Father who promises eternal happiness in the next life as compensation for enduring the hardships of this world. Life, for Whitman, is a series of "perpetual transfers and promotions"—an endless exchanging and evolving of elements. By reimagining the earthly realm as a world already infused with the divine, Whitman makes irrelevant the Christian need to be saved from its sinful entanglements. For Whitman, death is the transfiguring, shape-shifting part of creation, and time is an endless circling of change within the eternal now—and thus deathless:

> The smallest sprout shows there is really no death,
> And if ever there was it led toward life, and does not
>   wait at the end to arrest it. . . .

> All goes onward and outward . . . and nothing
>   collapses,
> And to die is different from what any one supposed,
>   and luckier.

Luckier. I can't help thinking that so much of my family's Christianity was based on fear—fear of death and the possibility of hell. One had to be good *or else*. It is obviously a strong incentive. But in the end, it feels impure, even childish. One isn't acting justly because that impulse comes from somewhere within. Rather, like a child, one does the right thing only out of fear of the repercussions of acting otherwise. But what if the promise of heaven and the prospect of hell were gone? With it would go the burden of guilt that bore down so hard on my father. What would remain? Perhaps a more affirmative and truly just religion that looks far more like the gospel of Thomas Jefferson and the poetry of Walt Whitman.

Near the end of my time at Furnace Mountain, I stopped by the teahouse, where Clay, one of Dae Gak's students, and his son Sam were lunching on tofu and cabbage. (Once I asked Dae Gak if he was a vegetarian; he replied, "I am when I'm eating vegetables.") Clay called out, "Here comes the hermit from the eastern mountain," which pleased me immensely. He offered me some lunch, but I said I was on my way up to a capstone that local maps call State Rock.

"I like to call it Vulture's Peak," Clay said.

"How come?" asked Sam.

"That was where the Buddha gave a famous sermon."

Sam giggled, incredulous.

Clay was imperturbable. "No, really," he said. "The Buddha was sitting at a place like that, and there were hundreds of people gathered around him. But he didn't say anything. He just sat there, and everyone started grumbling about how he was a fraud. Then he stood up and held out a flower. One of his students, Kasyapa, smiled. He understood. And that was the first transmission. Kasyapa became the first patriarch of Zen."

The climb to State Rock was steep, but not long. At the base of the trail sat a roughly hewn stone Buddha. Mayapple, with its single green leaf, grew around its base. Further up the trail, as the dirt path gave way to rock, I grabbed on to the exposed roots of chestnut oaks and pulled myself up over the sandstone boulders. When I reached the capstone that Clay had renamed, there actually were two turkey vultures circling nearby. Thrifty mountain laurel and shortleaf pines grew in the thin humus that spread across the back side of the capstone. Names, dating back to World War I, had been etched into the rock.

The day was sunny and breezy. I sat down and pulled *Leaves of Grass* out of my backpack. What Clay had said about the Buddha holding up the single flower reminded me of something similar—perhaps the image, or perhaps only the implication—in *Song of Myself*. I leaned my back against one of the stubby pine trunks and read until I found the line I half remembered: "A morning-glory at my window satisfies me more than the metaphysics of books." That must have been close to what the Buddha meant when he held up the one flower. To those eager to know the timeless answers, the unshakable truths, the poet and the prophet each offer up the most transient of images. Forget metaphysical speculation, they seem to be saying, which in the end is only speculation; instead, consider the lilies.

Like the Buddhist, Whitman also refuses to speak of what he cannot know, the meta-, and praises instead what he does—the physical. To know nothing about the nature of God, but to see that Presence within the natural world—some contemporary theologians have taken to calling this panentheism. The second syllable—the *en*—is important. This isn't, they say, the *worship* of nature—pantheism—but rather the recognition of the Creator's presence in the Creation. This idea became incredibly important to Emerson as he began defining his new American gospel, transcendentalism. "As the river flows," he wrote, "and the plant flows (or emits odors), and the sun flows (or radiates), and the mind is a stream of thoughts, so was the universe an emanation of God." If the universe *emanates* from God, then there is no separation between the Creator and the Creation. One is an extension of the other.

I found this idea thrilling, perhaps because it stood in such contrast to my grandfather's belief in our helpless alienation from God. Emerson proposed that when the individual perceives this divine extension, the self achieves a mystical unity with the One: "In thoughts *of* God or *about* God the Subject is separated from the Object, but in ecstatic union there is no such separation." Something startling happens here. To achieve this union of humankind with the natural world means to reverse the Fall of Man. How does one convey this sublime realization? For the Buddha and Whitman, by holding up a single flower. No words, no doctrine, no metaphysic. The flower, like the poetic image, calls the reader and the listener into a clearing where the only laws are the laws of one's own nature, which is a manifestation of the Original Nature that lies at the heart of Clay's Buddhism and American transcendentalism.

Walt Whitman claimed to be a *kosmos* because he thought the laws of the universe were written in the unspoken intuitions of one's own mind. That is to say, for Whitman, we are at once individual cells within the vast macrocosm of the universe, as well as individual versions, microcosms, of that great Unity. Thus he was at once a spiral nebula—"The whirling and whirling is elemental within me"—and a loafing poet contemplating a blade of grass. William James praised such high thoughts as "the sublimest achievement of intellectualist philosophy," but he finally put them aside as purely "mystical," theories that could not be tested by science. However, in 1868, Charles Darwin was already suspecting that there was more than just an elegant metaphor at work here. "Each living creature must be looked at as a microcosm," he wrote, "a little universe, formed of a host of self-propagating organisms, inconceivably minute and as numerous as the stars of heaven."

More recently, the brilliant American bacteriologist Lynn Margulis has proposed that every *kosmos*, from the microcosmic organism on up to our macrocosmic planet, has evolved through a very long game of symbiosis, "the coming together that leads to physical interdependence and the permanent sharing of cells and bodies." Two billion years ago, free-floating bacteria combined to form what Margulis calls "bacterial confederacies." These confederacies gradually formed a thin membrane that held them together. Then some of the bacteria turned into the oxygen-using mitochondria, and a command center, the nucleus, took shape. The cell was born, or rather, self-made through what some scientists today call *autopoesis*. Then cells took up residence inside larger organisms that in turn developed their own protective membranes. And they are still with us. "The descendants of the bacteria that swam in primeval seas breathing oxygen three billion years ago exist now in our bodies as mitochondria," writes

Margulis in her book *Microcosmos*. The larger flora and fauna in turn formed communities within forests and other ecosystems that also acted as even larger, self-regulating organisms. And all of these biomes are finally protected by an even larger membrane, the atmosphere. Margulis refers to her theory, her scientific creation story, as *symbiogenesis*. As the ultra-Darwinist and world-class skeptic Richard Dawkins has noted, "Not only is Dr. Margulis's theory of origins—the cell as an enclosed garden of bacteria—incomparably more inspiring, exciting and uplifting than the story of the Garden of Eden, it has the additional advantage of being almost certainly true." Indeed, the poetry of Margulis's theory is stunning when one considers how the macrocosmic planet and the microcosmic cell reflect such parallel patterns of self-creation. And in between these two micro- and macroscopic universes stood the human, fleshy *kosmos* that was Walt Whitman.

Margulis has likened the image of the earth we first glimpsed in photographs taken from space to the image that Narcissus first glimpsed in the water. Only this time, the earth resembles a single cell—another microcosm within the vast universe. Narcissus realizes that every cell that makes up his own being is reflected in the image of a small blue planet. He looks at the earth and sees that it is one self-regulating organism that, like the human body, needs its symbiotic parts.

No one better personifies this modern version of Narcissus than Whitman. As he winds up for the final sections of *Song of Myself*, Whitman recounts in a wild tableau the astral, evolutionary journey of his own becoming:

> Before I was born out of my mother generations
>   guided me,
> My embryo has never been torpid . . . nothing could
>   overlay it;
> For it the nebula cohered to an orb . . . the long slow
>   strata piled to rest on . . . vast vegetables gave it
>   substance,
> Monstrous sauroids transported it in their mouths and
>   deposited it with care.
>
> All forces have been steadily employed to complete
>   and delight me,
> Now I stand on this spot with my soul.

From carbon strata to dinosaurlike "sauroids," the passage suggests that Whitman, like Margulis, understood the *self* as a vast accumulation of deeply historical, mutually dependent alliances. Whitman felt intuitively what Margulis affirms scientifically: "Our bodies contain a veritable history of life on

Earth. Our cells maintain an environment that is carbon- and hydrogen-rich, like that of the Earth when life began. They live in a medium of waters and salts like the composition of the early seas." It is thus in a very real sense that Whitman can claim, in *Song of Myself*, "Sea of the brine of life! / I am integral to you. . . . I too am of one phase and of all phases." He believed that everything is simultaneously in the process of evolving into another version of the *kosmos* and dissolving back into that larger being, that larger self. Margulis's theory of symbiogenesis also recalls Whitman's understanding of the self as an exchange of atoms—"For every atom belonging to me as good belongs to you." Margulis admits that her theory "violates our view of ourselves as discrete physical beings separated from the rest of nature," but in doing so, it offers a scientific foundation to Whitman's natural philosophy—the self is realized simultaneously in what lies beyond its own walls and in what dwells deep within its core. That finding and forgetting the self are ultimately the same thing—this is also one of Whitman's great lessons. To forget the small, individual self is to find the expansive, orderly, interdependent self that Whitman called the *kosmos*. As Emerson famously remarked in his first book, *Nature*: "Standing on the bare ground,—my head bathed by the blithe air, and uplifted into infinite space,—all mean egotism vanishes. I become a transparent eye-ball; I am nothing; I see all; the current and the Universal Being circulate through me, I am part or particle of God." The small self, the mean ego, is lost in the finding of the Universal Being. Of this larger Self, Whitman wrote near the end of *Song of Myself*:

> It is not chaos or death . . . it is form and union and
>   plan . . . it is eternal life . . . it is happiness.

Such a "plan" is not, perhaps, the predetermined deistic model of Thomas Jefferson's Creator. Whitman said he accepted Darwin's theories of evolution "from A to izzard." Rather, Whitman seemed to understand the *kosmos* in the same way as Alexander von Humboldt, the far-ranging naturalist who popularized the term in Whitman's lifetime. For Humboldt, the *kosmos* was "one great whole animated by the breath of life."

Since Humboldt, this idea has remained on the periphery of the scientific lexicon, often taking new names, supported by new experiments. The influential mathematician and philosopher Alfred North Whitehead argued in the 1920s that science needed to replace a materialist, mechanistic view of nature with one that emphasized "the organic unity of a whole." He posited, as would Margulis, that the elements of the natural world function as an organic, densely textured web. Moreover, argued Whitehead, philosophers and scientists should listen to poets like Wordsworth and Whitman because it is they who best intuit that the particulars of the natural world are not mere matter but parts of a greater whole, organisms within larger organisms. Most recently, Margulis and James Lovelock, using harder science and computer models, have affirmed Whitehead's theories, showing that the planet Earth functions as a symbiotic, self-regulating, single organism. Margulis and Lovelock have demonstrated that as the amount of solar energy reaching the earth has increased over the last three billion years, the amount of heat-trapping carbon dioxide has decreased (which is to say, the earth seems to know what it is doing, even if the carbon-pumping humans that are creating global warming do not).

Whitman's twist on all of this was to proclaim that natural laws are reflected within one's own nature so that "whatever satisfies the soul is truth." This exuberance, this boldness, this self-*confidence*, is perhaps what his intellectual heir William James liked best in Whitman, what he thought might be an antidote for those like my father who suffered from what James called the "nightmare view of life." He imagined Whitman "loafing on the grass on some transparent summer morning," and "feeling the sufficiency of the present moment, of its absoluteness—this absence of all need to explain it, account for it or justify it." At that moment, Whitman (as channeled by James) realized, "To feel 'I *am* the truth' is to abolish the opposition between knowing and being." In this formula, truth is a product of the soul, and this truth rejects all abstract, conceptual knowledge so that the poet is returned to the direct experience of the moment. The poet, unlike the philosopher, does not need to *know* the truth; he or she wants to *enact* the truth. Just as Darwin showed how the *kosmos* is a self-creating, endlessly improvising entity, so the poet dramatizes throughout *Song of Myself* his own invention of "Walt Whitman, an American, one of the roughs, a kosmos."

There is one more critical confluence between Whitman the poet and Margulis the scientist—for both, biological symbiosis on the molecular level also offers a vital lesson in democracy. Margulis rejects Herbert Spencer's phrase "survival of the fittest" to explain Darwin's theory of natural selection or to justify social exploitation and cruelty. It should be clear enough to anyone who has watched footage of shoppers trampling one another at Wal-Mart on the day after Thanksgiving that consumer capitalism thrives on antisocial, hyperindividualistic behavior. But this is not the way we successfully evolved as a species. And it is not the way life evolved. Instead, Margulis understands "survival of the fittest" to mean not predatory violence but rather cooperative behavior that began on a bacterial level and eventually led to a species, *Homo sapiens*, that hunted large game together and shared food around a fire. "Life did not take over the globe by combat, but by networking," she counters. Darwin, in working out his theory of natural selection, realized that nature is constantly increasing its diversity as an alternative to straight-out competition for resources. Organisms changed and evolved to find new biological niches. Creativity, not confrontation, is often the name of the game. As Margulis writes in *Microcosmos*:

> Competition in which the strong wins has been given a good deal more press than cooperation. But certain superficially weak organisms have survived in the long run by being parts of collectives, while the so-called strong ones, never learning the trick of cooperation, have been dumped onto the scrap heap of evolutionary extinction.

Margulis, like Whitman, very clearly sees a democratic ethic at work in the evolutionary processes of the natural world. Life does not have to be a zero-sum game with winners and losers. In fact it might be a non-zero-sum game in which, through cooperation and reciprocation, many players win. The research that best reflects this optimism comes from the field of evolutionary biology. In 1971, Robert Trivers developed the theory of "reciprocal altruism"—you scratch my back, I scratch yours—and argued that it was a trait humans had learned as an evolutionary tool of survival. In the late 1970s, the American political scientist Robert Axelrod tested this theory by asking more than two hundred social and natural scientists to submit computer programs for a game called The Prisoner's Dilemma. The name stems from a scenario in which two prisoners are asked to give evidence against the other in exchange for a lesser sentence. If both remain silent, if neither "defects," then the police can convict them both only on the lesser charges. But the prisoners do not know this. If only one "defects," he will win and get the lesser sentence. Yet if each turns on the other, they will both get longer sentences. Axelrod converted this contest between cooperation and self-interest into a simple mathematical formula: In the experiment, "players" were given points for cooperating with each other or taking advantage of each other. Each player gets three points if they both cooperate and one point if they both defect; but if one defects while the other cooperates, the defector gets five points and the cooperator none. The game would thus seem to reward aggressive behavior, but the reverse proved true. Axelrod found the programs that survived the longest were those that emphasized cooperation over aggression. Cooperating players got tired of being taken advantage of, and they slowly isolated the defectors, who then received no points. If The Prisoner's Dilemma is a game of "survival of the fittest," then the "fittest" programs, those that ran the longest and earned the most points, proved to be programs that emphasized cooperation. Margulis, along with her son and coauthor Dorion Sagan, concluded, "Axelrod's work is consistent with our view that all large organisms came from smaller prokaryotes that together won a victory for cooperation, for the art of mutual living." It also suggests that Americans' hyperindividualism has pushed us into an economy and a culture that emphasizes competition for status and wealth at the expense of cooperation and community. But if cooperation is so imperative to biological communities, from cells to forests, we might do well to pay closer attention to nature's intelligence. It was, I think, in this spirit that Secretary of Agriculture Henry Wallace said in 1930, "There is as much need today for a Declaration of Interdependence as there was for a Declaration of Independence in 1776."

What is astonishing about *Song of Myself* is not only how often it anticipated contemporary science, but also how Whitman could hold together the seemingly contradictory ideas of the *one* and the *many* within his own philosophy, his vision of the Self. He could take in all of the diversity of the natural world and still intuit that it all began when a lightning bolt struck a single cell as the planet cooled billions of years ago. Then he could fathom how that microcosmic cell and the macrocosmic planet function by similar laws of organic symbiosis. Finally, he said, the unfolding of the natural world, as well as the unfolding of one's individual nature, is an infinite reflection of a primal law that says the Creation is the body of the Creator, and each organism within that Creation is a microcosm that embodies the same self-organizing principles of the original.

In a passage near the end of *Song of Myself,* Whitman summarizes his natural philosophy:

I have said that the soul is not more than the body,
And I have said that the body is not more than
   the soul,
And nothing, not God, is greater to one that
   one's-self is,
And whoever walks a furlong without sympathy walks
   to his own funeral, dressed in a shroud,
And I or you pocketless of a dime may purchase the
   pick of the earth,
And to glance with an eye or show a bean in its pod
   confounds the learning of all times,
And there is no trade or employment but the young
   man following it may become a hero,
And there is no object so soft but it makes a hub for
   the wheeled universe,

And any man or woman shall stand cool and
    supercilious before a million universes.

The duality between body and soul has disappeared. One
need not worship a historical or heavenly God, because all
men and women possess the divine spark that is suffused
throughout the Creation. All occupations might render men
and women heroes. Just as quantum theory has shown that
one's version of reality depends on where one stands to ob-
serve it, and just as ecology has shown that all organisms are
crucial elements in the greater whole, so Whitman declares
everything to be at once the center of a universe, each as important as the other. Here at the end of his poem, Whitman
returns his American folk hero to the eternal now—"You
must habit yourself to the dazzle of the light and of every
moment of your life." Here is at once a theology, a psychology, a politics, and a morality—all woven together to form a
radical definition of the Self. The free spirit, the soulful individual, abandons the laws of rigid orthodoxy. Morality is no
longer a matter of thou-shalt-nots. Rather, as the contemporary American philosopher Richard Rorty has written, "Moral
progress is a matter of wider and wider sympathy." It is the
opposite of doctrine and ideology. Within Whitman's idyllic
America, democracy depends on the possibility and variety

of true self-invention, along with true sympathy for the other selves that together create the body politic of the nation.

Near the end of his famous study *The Varieties of Religious Experience*, William James concluded, "Not God, but life, more life, a larger, richer, more satisfying life, is, in the last analysis, the end of religion. The love of life, at any and every level of development, is the religious impulse." It sounds like a perfect commentary on the poetry of Walt Whitman, yet quite foreign to my father's and my grandfather's particular variety of religious experience. I do remember times when my grandfather would be poking around the yard in the morning, raking leaves or filling the birdbaths, and he would be quietly holding a conversation with God. Often it was a low singing, something that sounded like an improvised hymn that praised the ordinary miracles of the Creation— the sparrows, the willow trees, the mild morning sun. Yet many other times, my grandfather's religion quickly shed this Jamesian affinity for life and became instead an indictment of this world where humankind deserves death and the pain delivered upon us for disobeying God back in that far-off garden. Toward the end of my grandfather's life, he watched my grandmother drift into dementia while his relationship with my uncle—which had been strained since my father's death—gradually devolved into an irreconcilable mutual reproach. During those years, I watched my grandfather gradually come to hate his life, to hate the world, and to wish he possessed his son's "courage" to take himself out of it.

I can't say whether I agree with my grandfather that my father's suicide was a courageous act. In the end, I don't know enough to pass judgment, nor do I feel the inclination. But I have reached the conclusion that the suicide is someone overcome by forces—internal and external—that seem so averse to his or her own nature that life itself seems no longer worth the struggle. My father was combating both a chemical imbalance within his brain and my family's oppressive faith, which was always there to remind him of his inferiority, his abiding sinfulness. Combined, those forces were simply too much for him. In the essay "Is Life Worth Living?" William James proposed that what does makes life worthwhile is a belief in something like Emerson's Over-soul, a belief that "behind nature there is a spirit whose expression nature is." Such a belief rejects the orthodoxy of established religion. Rather, "to trust our religious demands means first of all to live in the light of them, and to act as if the invisible world which they suggest were real. It is a fact of human nature, that men can live and die by the help of a sort of faith that goes without a single dogma or definition." This religion freed from doctrine and dogma was, of course, Whitman's faith. It was the reason he could charge his readers:

. . . Be not curious about God,
For I who am curious about each am not curious
   about God,
No array of terms can say how much I am at peace
   about God and about death.

I hear and behold God in every object, yet I
   understand God not in the least,
Nor do I understand who there can be more
   wonderful than myself.

Anyone who does not wish to see God "better than this day" is certainly in no danger of dying by suicide. Nor is anyone who, like Whitman, feels such a Jamesian assuredness that the divine lurks behind all things, though we can only intuit it. James himself concluded: "This life *is* worth living, we can say, *since it is what we make it, from the moral point of view*; and we are determined to make it from that point of view, so far as we have anything to do with it, a success." One can invent one's own life, one's own self, *outside the Law*, driven by one's own inner convictions, one's own will to master the circumstances of one's own life. One might, like Whitman, set one's own stage, invent one's own role in the existential drama, and call it a success. But this is, of course, exactly what my father could not do. He could not define himself outside the Law set down by my grandfather.

For many years I myself carried, if not my father's sins, at least his sense of his own sinfulness, and mine, before the God of my grandfather. And when I reached the age at which my father ended his life, my own defenses started collapsing. I had gone to Furnace Mountain in hopes that the quiet, the meditation, and the woods would do me some good, restore my balance, perhaps even purge some of my family's demons. I can't say that I left Furnace Mountain a brand-new man. But that summer did set me on a new course. My anxiety attacks stopped, as did the migraines. And at Furnace Mountain, I did come to realize that James was right: Life, not the Christian promise of life after death, is the purest religious impulse. Wandering and loafing along those ridges, I began to shed the blinding mental apparatus that says the world is fallen, that we are born culpable of some far-off transgression, that human suffering is our punishment for that sin. Nothing as beautiful as the broadleaf forest seen from Vulture's Peak could be fallen, tainted by my sinfulness.

"Our resistance to the wilderness has been too strong," wrote William Carlos Williams. "It has turned us anti-American, anti-literature." As I noted earlier, Williams blamed the Puritan legacy for driving this theological wedge between the Creator and the Creation. Unfortunately, many American corporations moved to fill that gap, and Andrew Carnegie's gospel of wealth came to coexist far too easily alongside the biblical Gospels. The Sermon on the Mount gave way to Carnegie's "law of competition," which said that "while the law may be sometimes hard for the individual, it is best for the race, because it insures the survival of the fittest in every department." Whitman's democratic vision of common, heroic men and women gave way to Carnegie and his contemporary heirs who raided the commons for personal gain.

In a late manifesto, "Democratic Vistas," Whitman cast a gaze over the nation and found "depravity of the business classes," "a mob of fashionably dress'd speculators and vulgarians," and cities that "reek with respectable as much as non-respectable robbery and scoundrelism." The country's "unprecedented materialistic advancement" had created a nation that was "canker'd, crude, superstitious and rotten." In short, American democracy was "an almost complete failure in its social aspects, and in really grand religious, moral, literary and aesthetic results." America had failed—as it still fails—to live up to its virtuous depiction in *Song of Myself.*

But, said Whitman, all was not lost, because true American democracy lay in the future. "Time is ample," he wrote. "Let the victors come after us." Democracy remains "a great word, whose history, I suppose, remains unwritten, because that history has yet to be enacted. It is, in some sort, younger brother of another great and often-used word, Nature." The greatest lessons of nature, claimed Whitman, were variety and freedom, and Americans had not learned them. Solidarity had been displaced by a sad conformity. Religion remained a set of creeds and conventions, not a force that inflamed the soul. Thus, because it had not yet learned the lessons of nature, "America has yet morally and artistically originated nothing." It's quite a charge. American culture was weak and corrupt because it came out of "parlors or lecture rooms," not from "the mountain peaks, the ocean, and the tumbling gorgeousness of the clouds." Whitman believed, very much like Thomas Jefferson, that morality, religion, democracy, and art all found their true roots in nature. In "Democratic Vistas," he sought one last time to turn future Americans back toward the populist, agrarian version that Thomas Jefferson urged upon the nation.

That America lay, and still lies, in the future. Whitman believed we needed a great poem and a new religion, a natural religion, to get us there. Whitman wrote the poem, and he called it *Leaves of Grass*. That one hundred years later a group of American New Testament scholars would unearth a new Gospel that set out just such a natural religion is as astonishing as it was unlikely. But as the next chapter demonstrates, that is exactly what happened.

III

The Kingdom of God

Years after I gave up on the church, I finally discovered a Christianity that I could accept. I found it in a manuscript that had been lost to New Testament scholars for thousands of years—the Gospel of Thomas. But more important to the intellectual and spiritual lineage I have been tracing throughout this book, I found a biblical gospel that validates the American gospel of William Byrd, Thomas Jefferson, Ralph Waldo Emerson, and Walt Whitman. And because the Gospel of Thomas predates the four canonical Gospels of Matthew, Mark, Luke, and John, it offers what is at once a more authentic version of Jesus' teachings and a more authentic American gospel.

I have made the case throughout this book that there is an

alternative, more authentic American gospel than the one championed by the mainstream church, which has replaced the teachings of Jesus with those of the apostle Paul. It is a gospel that can be found, in compromised forms, in the canonical New Testament. But to find its purest form, we must turn to the Gospel of Thomas. "Thomas" comes from the Semitic word for "twin." That Thomas Jefferson's version of Christianity actually found a twin gospel—one that, like his, included no miracles, no claims of divinity, but only the teachings of Jesus—hidden beneath an Egyptian cliff, and that this ancient gospel was also recorded by a man known as Thomas, makes for a remarkable story.

Sometime near the end of the nineteenth century, two British archaeologists, Bernard Grenfell and Arthur S. Hunt, were searching through an ancient trash heap along the Nile River, at a site known as Oxyrhynchus, when they found three small papyrus leaves that seemed to come from the same Greek codex. One of the fragments read, "These are the . . . sayings . . . the living Jesus spoke and . . . also called Thomas." New Testament scholars have long known that there once existed a Gospel of Thomas, because in the third century, Hippolytus denounced such a text in his *Refutation of All Heresies.* And because Thomas's gospel ran afoul of the early church bishops, particularly Irenaeus, most copies of it were likely destroyed. Grenfell and Hunt dated their findings to around 200 CE and published the fragments in 1897 as "Sayings of Jesus."

Fifty years later, 150 miles downstream in another river town, Nag Hammadi, an Egyptian farmer named Muhammad Ali al-Samman was guiding his camel beneath the nitrogen-rich cliffs that line the Nile, collecting peat fertilizer for his fields. As he dug at the base of one cliff, Muhammad Ali unearthed a sealed red jug, obviously ancient. Fearing a jinn but hoping for gold, he broke the jar open with his mattock. He found neither. All that fell out were thirteen books (codices), made from papyrus and bound in leather. Figuring the manuscripts might at least be worth something, Muhammad Ali gathered them up in his turban and carried them home. According to the New Testament scholar James A. Robinson, who has pieced this whole story together, Muhammad Ali's mother used some of them to ignite their outdoor clay oven. Muhammad Ali traded others for oranges and cigarettes.

Meanwhile, shortly after the discovery, Muhammad Ali and his brothers hacked to death a man they claimed had killed their father twenty years earlier. The brothers cut out the dead man's heart and ate it so as to put their father's spirit at ease. But when local police started poking around, asking about the murder, Muhammad Ali didn't want to risk any more trouble over the codices he had found in the cave. Since the manuscripts were written in Coptic, the variant of Greek used by the Copts, Egyptian Christians, he hid one at the house of a Coptic priest. The priest, in turn, sent it to Cairo by way of his brother-in-law to ascertain its value on the antiquities market. But someone tipped off Egyptian authorities, who then took the brother-in-law into custody and told him he could return home only if he made a "gift" of the codex to Cairo's Coptic Museum, which he promptly did.

Here a notorious one-eyed bandit named Bahij Ali enters the story. Cairo's leading antiquities dealer, a Cypriot named Phocion J. Tano, had hired Bahij Ali to bring to bear his considerable powers to persuade Muhammad Ali to sell him all of the remaining codices for a nominal price. But again, the Egyptian government heard about Tano's acquisition and pressed him to entrust the manuscripts to the Coptic Museum for "safe keeping." Tano spent most of the 1950s trying unsuccessfully to get the codices back.

In 1952 the French scholar Henri-Charles Puech realized that a tractate in Codex II contained sayings that matched the Oxyrhynchus fragments. Less than sixty years after Grenfell and Hunt uncovered hard evidence that a Gospel of Thomas did at one time exist, Puech was able to conclude that the entire text had been found.

When all of the remaining codices where accounted for, there turned out to be fifty-two separate tractates hidden at

Nag Hammadi. Scholars have puzzled over how these texts ever ended up in this remote port town. I think Elaine Pagels offers the most probable explanation: In 325 CE, the Roman emperor Constantine, newly converted to Christianity, called for a conference of bishops in Nicea. He was frustrated that so many people who called themselves Christians seemed to follow so many different doctrines. At Nicea, he charged the bishops to come up with a short document that would unite Christians and eradicate heresy. Years earlier, Irenaeus, the bishop of Lyons, had argued that the Gospel of John should be part of the New Testament canon because, though it differed wildly from Matthew, Mark, and Luke, it was the only Gospel that actually asserted the divinity of Jesus. (Solely in John's gospel does Jesus claim, "I am the way, the truth, and the life. No man comes to the father except by me." John 14:6, NKJV.) And this message—this Christology—so different from that of the other Gospels, dominated the conference at Nicea. Irenaeus's student Athanasius, the forceful bishop of Alexandria, shaped the new document so it closely resembled the Christ claims in John's gospel. As a result, the Nicene Creed reads as if Jesus had never delivered the Sermon on the Mount. But though it contains nothing of Jesus' teachings, the document apparently pleased Constantine, who pressured all of the bishops to sign it.

Forty-two years later, Athanasius issued his annual Easter

letter in which he offered a list—the first known list—of what were the twenty-seven acceptable books of the New Testament Bible. Then he called for all other "heretical" documents to be destroyed. However, scholars now suspect that a recalcitrant monk at the Saint Pachomius monastery, near Nag Hammadi, refused the order and instead buried the codices in a large red jug, where they stayed hidden until Muhammad Ali al-Samman accidentally dug them up in 1945.

Unfortunately, years of infighting among international scholars stalled the publication of what came to be called the Nag Hammadi Library. To make matters worse, the European countries that owned the publication rights showed a remarkable indifference to the task. In the end, an American, James M. Robinson, obtained photocopies of the individual Coptic tractates throughout the 1970s and passed them on to a team of American translators. As a result, the first definitive edition of the Nag Hammadi Library was published in English.

Perhaps because of this head start, much of the groundbreaking scholarship devoted to the Gospel of Thomas has come from Americans, namely Robinson himself, Stephen Patterson, Helmut Koester, Robert Funk, Stevan Davies, and Elaine Pagels. It is as if Thomas Jefferson's *Life and Morals of Jesus of Nazareth* had prepared the Americans for what they would find in the ancient Gospel of Thomas. In some

Borgesian way, Jefferson's gospel has become a predecessor to the Gospel of Thomas, though it was composed some 1,700 years later.

The similarities between the two gospels are remarkable, as much for what they do not say as for what they do. Like Jefferson's gospel, Thomas's ignores the virgin birth. Thomas's Jesus never performs a miracle, never calls himself the Son of God, and never claims that he will have to die for the sins of humankind. Instead he tells parables, he issues instructions, and most alarmingly, he locates the kingdom of God in that one place we might never look—right in front of us.

On the topics of sin, sacrifice, and salvation, Thomas's Jesus, like Jefferson's, is silent. In fact, what we find in the Gospel of Thomas looks very different from what we find in mainstream Christianity. It predates the church, it predates the synoptic Gospels (Matthew, Mark, Luke), and it predates any claim, by Jesus or his followers, that he is the Christ. There is no attempt in the Gospel of Thomas to tell the "story" of Jesus, and there certainly is no inkling of some impending Day of Judgment. Instead, Thomas offers a "sayings gospel," a collection of 114 apposite responses Jesus is remembered to have made to questions posed by his followers and anonymous crowds. These were compiled under the name of Thomas and were circulated throughout southern Syria among a group that scholars now call the "Jesus movement."

As a literary type, the Gospel of Thomas bears kinship with the "wisdom collections" of late Judaism, such as Proverbs, Ecclesiastes, and the Wisdom of Solomon. But its closest counterpart is the sayings gospel Q (after the German *Quelle*, or source), from which Matthew and Luke took much of their material. Many of the sayings in Thomas's gospel also appear in Q; almost half of the sayings in Thomas would be familiar to any reader of the New Testament. Such collections circulated around Hellenistic Palestine simply because they contained advice people wanted to remember.

I am not a New Testament scholar. All I bring to the debate is a childhood of compulsory churchgoing and a master's degree in reading literary texts. Yet it seems I have spent the eighteen years since I stopped going to my parents' Baptist church attempting to extricate myself from its hold on me. In doing so, I learned a lot about the various groups who were struggling in the early part of the first century to understand exactly what to make of this Mediterranean wanderer named Yeshua, whom today we call Jesus. I learned how the four canonical Gospels were put together, what sources they used, and which words were more likely uttered by the speaking Jesus. And I have reached the conclusion that the Gospel of Thomas offers an earlier, very different portrait of Jesus, one more authentic than those found in Matthew, Mark, Luke, or John. Moreover, I think the Gospel of Thomas could and should offer a profound corrective to mainstream Christianity as it is practiced in the United States.

Before I explain why, I should provide some context for such claims. Over the past two thousand years, the most important document, the most important story for the western hemisphere, and much of the southern, has been the Gospel According to Mark. This narrative of the Son of God's intervention in this world, his ability to work all manner of miracles, his promise to return again and establish a kingdom of God, and his martyrdom "as a ransom for many" has fired European and American imaginations like no other religious story. Though Mark's may be the clumsiest of the Gospel narratives, it is the oldest, written in the early seventies CE. We know this because Matthew and Luke borrowed extensively from it, while John ignored it altogether. When the language in Luke and Matthew is similar or identical, it mirrors a passage from Mark, or the sayings gospel of Q. In the 1970s and 1980s, the Q manuscript was discovered, or perhaps more appropriately, it emerged, embedded in the Gospels of Matthew and Luke. That is to say, New Testament scholars noticed that there were striking similarities in the wording of Jesus' sayings that do not appear in Mark. Some scholars argued that Q could not have existed, because there were no other examples of a Gospel that was only a collection of Jesus' sayings. But the discovery of Thomas's gospel has silenced that objection. As in the Gospel of Thomas, many of the *Q* sayings have to do with that mysterious phrase "the kingdom of God." And as in Thomas, and in contrast to Mark, the Jesus of *Q* (at least in the earliest versions of *Q*) refused to locate that kingdom in another, heavenly realm. Rather, he asserts that the kingdom of God is already "in your midst."

Mark's gospel, on the other hand, made a radical departure from this understanding of the kingdom, and the reason for it tells us something very significant about early Christianity. Mark (or whoever wrote the Gospel) composed his narrative shortly after the Jewish–Roman War (68–72 CE), in which the Romans razed the Second Temple and brought the Jerusalem temple state to an end. The logic of his gospel hinges on this fact. The story of Mark is in many ways the story of a failed mission. A savior came into the world, but he couldn't save it. He gathered around him a group of followers who turned out to be incredibly obtuse; even the simplest parable seemed beyond them. And before he could convince his own people that he was their Messiah, they betrayed him and he was crucified. How could this have happened? Burton Mack argues in *A Myth of Innocence: Mark and Christian Origins* that the Roman sacking of the Temple gave the author of the Gospel of Mark a way to rationalize Jesus' failure. Mack claims that Mark rigged his story so that Jesus actually prophesies the destruction of the Temple as punishment for those who do not acknowledge that he was the Son of God. Then Jesus prophesies that he will return to pass judgment of eternal damnation on those who do not bear witness to his divinity. Only then will he invite his believers into the eternal kingdom of God. *Next time it will be different,* he promises. For Mark, the destruction of the Second Temple and the execution of Jesus marked the end of Judaism and the beginning of a new religion. "By locating the Christ myth precisely as an originary event complete with social historical motivation and consequence, Mark created a story that was to give the Christian imagination its sense of a radical and dramatic origin in time," writes Mack. Out of a failed mission came the idea that such a failure was important. In fact, it was the beginning of something far more important. Everything hinged on the Crucifixion, on a martyr's blood sacrifice. And as the Puritans made clear enough, that feeling of uniqueness and superiority on the part of Christians has caused and justified all manner of cruelty toward the rival monotheisms and toward indigenous people, and this vindictiveness can be traced back to the Gospel of Mark.

But what if this story was wrong? Or what if it was not the whole story? Or was not the only story? What if there were followers of Jesus who did not buy into the apocalyptic logic of Mark? Or who did not build their communities around the saving sacrifice of Jesus' death, as did Paul? The last few decades of New Testament scholarship have shown that there was indeed such a group of followers. The persistent similarities between Thomas and Q—of the seventy-nine sayings of Thomas that appear in the synoptic Gospels, forty-six have parallels in Q—suggest that these early followers of Jesus were interested only in the wandering sage and the often strange things he had to say. This early group is the "Jesus movement" mentioned earlier, as opposed to the "Christ church" that would form later, and the radical difference between the two hinges on that ambiguous phrase "the kingdom of God." The Jesus movement was most likely made up of an itinerant people, like their leader, who moved on the fringes of both Jewish and Roman society. What we know from *all* of the surviving Gospels is that the early followers were mostly outcasts and day laborers, "foxes without dens." Yet they believed, on some level, that their kingdom, the kingdom of God, could replace the kingdom of Augustus Caesar. Mack nicely sums up the scholarly consensus about the Jesus movement as opposed to the Christ church: "Instead of people meeting to worship a risen Christ, as in the Pauline congregations, or worrying about what it meant to be a follower of a martyr, as in the Markan community, the people of Q [and Thomas] were fully preoccupied with questions about the kingdom of God in the present and the behavior required if one took it seriously."

One of the foremost Gospel of Thomas scholars, Stevan Davies, introduces his translation of it with this proclamation: "For those interested in Jesus of Nazareth and the origins of Christianity, the Gospel of Thomas is the most important manuscript discovery ever made." I think Davies is right for two reasons. Thomas's gospel can be dated more closely to the talking Jesus than can the canonical Gospels (more later on this), and the veracity of its sayings can be tested by the fact that many of them were recorded in other independent documents, mostly Mark and Q. What also bears repeating is that the Gospel of Thomas is the most important Christian manuscript ever discovered because its contents vary so strikingly from the twenty-seven books of the New Testament.

Unlike the newly discovered Gospel of Judas, Thomas's gospel is unsettling to many not because of what it adds to the story of Jesus but because of what it leaves out. The most unsettling omission for most Christians, and the reason Thomas's gospel will never be preached in churches like the one I grew up in, would have to be the omission of Jesus' claim that he is the Messiah, "the Son of Man seated at the right hand of power, and coming with the clouds of heaven," as Mark has it. Thomas's Jesus never calls himself the Messiah, the Christ, or the apocalyptic Son of Man, and as a result, he doesn't have to perform any miracles to prove his divine power. Many scholars have noted that both the Hebrew and the Aramaic translations of the phrase "son of man" would, in English, simply be "humankind." And John Dominic Crossan has further shown that even in the synoptic Gospels, when the apocalyptic Son of Man is mentioned, there are never two independent sources for that passage, which strongly suggests the phrase was one writer's own invention, a late addition from the Christ church. Thomas mentions the "son of man" once, in the same context as Luke and Matthew:

> Jesus said, "[Foxes have] their dens and birds have their nests, but the child of humankind has no place to lay his head and rest."*
>
> (Gospel of Thomas 86)

Here, the term clearly means "humankind" and, as will soon be evident, reflects the itinerant life of Jesus and his followers.

At one point in the Gospel of Thomas, a woman named Salome (perhaps the same Salome who in Mark's gospel discovers the empty tomb) directly puts to Jesus the question of who he thinks he is.

---

*All quotations from Thomas's gospel are from Marvin Meyer, trans., *The Gospel of Thomas: The Hidden Sayings of Jesus* (San Francisco: HarperSanFrancisco, 1992).

Jesus said, "Two will rest on a couch; one will die, one will live."

Salome said, "Who are you, mister? You have climbed onto my couch and eaten from my table as if you are from someone."

Jesus said to her, "I am the one who comes from what is whole. I was given from the things of my father."

"I am your follower."

"For this reason I say, if one is <whole>, one will be filled with light, but if one is divided, one will be filled with darkness."

(Gospel of Thomas 61)

This is one of the rare instances, however minimal, of scene-setting in the Gospel of Thomas. Jesus is eating at the table of a woman who is suddenly struck by this idea of *the wholeness of being*, enough so that she will give up her table, her home, her settled life, in order to find that wholeness. From this one scene, we can understand why Thomas's Jesus looks so different from the Jesus of the synoptic Gospels. His death bears no importance in Thomas because it isn't the martyred Jesus who will bring his followers into the kingdom of God; it is the living Jesus. And while Jesus claims to be a "son of God," in that he holds within him a divine light, that light isn't unique to him alone. As Emerson argued, anyone potentially might find it within him or herself. This Jesus does not have to die for his followers to find the kingdom of God. Nor does he have to offer his crucified body as a sacrifice that will earn their admission into that kingdom. *You can save yourself,* he tells the crowds. In fact, it is imperative: "Jesus said, 'If you bring forth what is within you, what you have will save you. If you do not have that within you, what you do not have within you [will] kill you.'"

What, then, are we left with? A Jesus who combines Thomas Jefferson's moral philosopher with Ralph Waldo Emerson's poet who is a liberating god. The advice in the Gospel of Thomas, like that in Jefferson's gospel, is extreme, so much so that the New Testament scholar Stephen Patterson has labeled it "counter-cultural wisdom." Like the Jesus in Jefferson's gospel, Thomas's Jesus is proposing a world turned upside down. He is especially hard on the rich. As in the canonical Gospels, he says that a man cannot serve two masters, and the poor will be the first to find the kingdom of God. He warns against lending with interest. He tells the parable of the rich man whose friends were too preoccupied with financial matters to come to dinner, and so he finally sent his servants out to "bring back whoever you find." As in the synoptic Gospels, the dinner party functions here as a symbol of radical equanimity—everyone, regardless of their social rank, is invited. But those preoccupied with making money have no time for, and no desire for, such equality and fraternity. And so, unlike Matthew and Luke, Thomas ends his story with this damning line: "Buyers and merchants [will] not enter the places of my father." As in the synoptic Gospels, the dinner party to which all are invited becomes the central image—the central enactment—of Jesus' message. Stephen Patterson has pointed out that most of Jesus' Sermon on the Plain, from Luke, also appears in Thomas, including the beatitudes about the poor, the hungry, the outcasts. By including all men and women at his table, Jesus absolved them of their social stigmas. These who have been shunned by "the world" are the ones Jesus invites to a table of equals in the eyes of God. Those who have been kicked aside by the Roman empire are welcome within the kingdom that he describes.

Thomas's Jesus also has no time for empty ceremony, such as fasting and praying. On the subject of circumcision, he points out, quite sensibly it seems to me, "If [circumcision] were useful, children's fathers would produce them already circumcised from their mothers." In addition, this Jesus is not at all concerned about sins of the flesh. "Why wash the outside of the cup?" he asks. "Do you not understand that the one who made the inside is also the one who made the outside?" And in an even more telling passage, he remarks,

"When you strip without being ashamed and you take your clothes and put them under your feet like little children and trample them, then [you] will see the child of the living one and you will not be afraid." The word "sin" appears only once in the Gospel of Thomas. And ironically, the sin is not gluttony but its ascetic opposite—fasting. Such then are the leveling politics of Thomas's Jesus. They are Whitmanesque in every sense—comradely, convivial, and classless.

Thomas's theology is not so easy to pin down. Thomas tells us in the first line of his gospel's prologue that what follows are the "hidden sayings" of Jesus. That is to say, their meaning is often esoteric and meant to spark some sudden, prerational realization on the part of the listeners. Then Thomas himself adds, "Whoever discovers the interpretation of these sayings will not taste death." I'll return to what that might possibly mean. What follows next are the first sayings of Jesus himself:

Jesus said, "Let one who seeks not stop seeking until one finds. When one finds, one will be troubled. When one is troubled, one will marvel and will rule over all."

Jesus said, "If your leaders say to you, 'Look, the kingdom is in the heaven,' then the birds of heaven will precede you. If they say to you, 'It is in the sea,' then the fish will precede you. Rather, the kingdom is inside you and it is outside you.

"When you know yourselves, then you will be known, and you will understand that you are children of the living father. But if you do not know yourselves, then you dwell in poverty, and you are poverty."

(Gospel of Thomas 1–3)

The kingdom is inside you and it is outside you—we recognize in that phrase a variant of Luke 17:21. But scholars have puzzled over whether Luke's Greek should be translated as "the kingdom of God is within you" or "the kingdom of God is in your midst." Thomas leaves no doubt. It is both, but it is the kingdom within that allows us to see the natural world as the kingdom of God. Or rather, each transforms the other. As in Whitman's *Song of Myself*, the kingdom within reveals the immanent divine nature of the world, and the kingdom outside affirms, as in Whitman's poetry, that human beings carry an inner reflection of that kingdom.

From the beginning to the end of the Gospel of Thomas, Jesus openly mocks the idea of some pie-in-the-sky, future kingdom of God. His followers are clearly alarmed by John the Baptist's vision of an impending apocalypse, at which time the Ultimate Arbiter will hear all grievances and right all wrongs. When they ask, "When will the rest for the dead take place, and when will the new world come?" Jesus replies, "What you look for has come, but you do not see it." *It's right in front of your face,* he tells them over and over, but "you do not know how to examine the present moment." Back at Furnace Mountain, when I asked Dae Gak for his thoughts on death, he wouldn't give me a straight answer. He said that if one lives wholly within the present, there is no past and no future, and therefore no death. I think there may be something of that sentiment in this passage from Thomas as well.

The main thing that prevents us from seeing the kingdom spread out before us is fear, namely the fear of death. Jesus refuses to talk about his own death in Thomas because he does not see it as a saving event, a cosmic interruption of history. When his followers do ask about their own deaths, he chides them for completely misunderstanding his teaching:

> The followers said to Jesus, "Tell us how our end will be."
>
> Jesus said, "Have you discovered the beginning, then, so that you are seeking the end? For where the beginning is, the end will be. Fortunate is one who stands at the beginning: That one will know the end and will not taste death."

Jesus said, "Fortunate is one who came into being before coming into being."

(Gospel of Thomas 17–18)

Jesus doesn't say that we won't die, only that we won't *taste* death. To live in a present that recovers the prelapsarian innocence of the beginning, to "stand at the beginning," means to shed the fear of death that only overtook the First Parents *east* of Eden. To claim there is no difference between the beginning and the end is, in effect, to negate the belief in a history that is marching toward an apocalyptic destination. Like Whitman, Jesus rejects a linear, millennial view of history, of time, for a passage back, as Whitman wrote,

> To reason's early paradise,
> Back, back to wisdom's birth, to innocent intuitions,
> Again with fair creation.

Jesus, throughout the Gospel of Thomas, is telling his followers to see the world again *as* the early paradise, what Whitman beautifully called a "realm of budding bibles." The cycles of the seasons return us again and again to another beginning. Accept the laws of nature as the laws of God, he tells them. As in nature, the end is another beginning. When Jesus prepares to send his followers out into the world to preach his message, he offers these instructions:

> If they say to you, "Where have you come from?" say to them, "We have come from the light, from the place where the light came into being by itself, established [itself], and appeared in their image." If they say to you, "Is it you?" say, "We are its children, and we are the chosen of the living father." If they ask you, "What is the evidence of your father in you?" say, "It is motion and rest."
>
> (Gospel of Thomas 50)

The natural laws of motion and rest are all the proof we need to know that we all "come from the light" that stands at the beginning. Another Thomas, Thomas Merton, once told an audience, "Life is this simple. We are living in a world that is absolutely transparent, and God is shining through it all the time." In the Gospel of Thomas, Jesus repeatedly speaks of this light within. This spark is proof of our kinship to the Creator—of our own divine beginnings. But human vanities blind us to it. We walk around wearing all sorts of lampshades until we finally convince ourselves that such a light never existed at all. The Jesus of Thomas's gospel is simply trying to give us back something we already possess. Here is a crucial passage:

Jesus said, "Images are visible to people, but the light within them is hidden in the image of the father's light. He will be disclosed, but his image is hidden by his light."

Jesus said, "When you see your likeness, you are happy. But when you see your images that came into being before you and that neither die nor become visible, how much you will bear!"

(Gospel of Thomas 83–84)

There is an empirical way of knowing, and there is an intuitive, Emersonian way of understanding. The "father's light" exists within everyone and "will be disclosed," but we cannot know it intellectually; we cannot give it a shape, an image. Likewise, we comprise two selves: the one we see in the mirror, and the face we had before we were born. It was that strange image of the original face that first drew me to Buddhism. Here it is in the Gospel of Thomas, where to "see" this imageless image, to know this original self, is to arrive at a nexus where the light within illuminates the world without, and finally shows it for what it truly is—the kingdom of God. And when Jesus' followers ask *when* they will enter the kingdom of God, he replies, "When you make the two into one, and when you make the inner like the outer and the outer like the inner, and the upper like the lower, and when you make male and female into a single one . . . then you will enter [the kingdom]." This idea of making two into one is central to the theology of Thomas. The Jesus of Thomas rejects the verbal and psychological dualisms at the heart of Paul's Christianity. Rather, like Zen Buddhists, Thomas's Jesus believes that to divide the world up into abstract categories is to miss actually seeing the world *as it is.*[*] At one point Jesus tells his followers, "On the day when you were one, you became two. But when you become two, what will you do?" When we come into being, Jesus seems to be saying, we are necessarily separated from the Creator, the One. What then? Jesus' question implies that we must rediscover the One, not by a return to some heavenly realm, but by recognizing that the world before us is an emanation of that One—an immanent wholeness, a kingdom unto itself.

The Jesus of both the Gospel of Thomas and the synoptic Gospels is, more than anything else, a poet—a man who spoke in concrete images and compelling parables. Like Homer, he was a poet of the oral tradition. Jesus never wrote anything down, probably because he couldn't. As John Dominic Crossan has pointed out, Jesus was almost certainly il-

---

[*]Richard Baker, the roshi of the Zen Center in San Francisco, once joked to Elaine Pagels, "Had I known the Gospel of Thomas, I wouldn't have had to become a Buddhist."

literate, as were ninety-seven percent of the Galilean artisan class to which he belonged. So instead of writing long doctrinal treatises like the scribes and Pharisees, Jesus spoke in stark images that his listeners would remember, and eventually write down. And because he did not teach from a written script, he spoke spontaneously and originally, revealing a native intelligence that quickly drew crowds. Because Galilee was an agrarian region, he focused on the imagery of seeds, plants, and food. But like all great poets, he did not tell his audience what his symbols or parables meant. That was up to them; they had to work for the meaning and, in doing so, make it their own. And throughout the Gospel of Thomas, Jesus' symbols bring back together the Creator, the Creation, and the Son of Man. This triad, what we might call a *realized trinity*, forms Thomas's kingdom of God.

To reach it, Thomas's Jesus offers succinct instructions: "Be passers-by." It's a phrase one might expect to find in the *Tao Te Ching*. Indeed, many scholars have noted the "Eastern" feel of Thomas's gospel. Edward Conze has even suggested that the Thomas Christians intermingled with Buddhists in southern India, and some scholars speculate that Jesus' "missing years" might have been spent in the East. I suspect it is the spirit of "nonattachment" in Thomas that seems so Taoist or Buddhist. It isn't a concept that acquisitive American Christians have ever been too comfortable with. Nevertheless, it is central to Thomas's gospel. His Jesus is trying to persuade "those who have ears" to shake off all of the world's distractions and encumbrances—family, reputation, money, temple—so they might finally *see something real*. It is the same impulse that drove Henry Thoreau out to Walden Pond. In his later essay "Walking," Thoreau echoes the dramatic call in Thomas to walk away from the world: "If you are ready to leave father and mother, and brother and sister, and wife and child and friends, and never seen them again,—if you have paid your debts, and made your will, and settled all your affairs, and are a free man, then you are ready for a walk." Thoreau is being purposefully dramatic here, to get his readers' attention, to remind them that they are not free, that they are bound to jobs and mortgages and commitments that rarely allow them to walk outside Concord. He even went so far as to invent his own etymology for "saunter." The saunterer, said Thoreau, was originally a *Saint-Terrer*, a Holy Lander who, "having no particular home, was at home everywhere." It is an excellent description of the Jesus movement that we find in the Gospel of Thomas. As an artisan by trade, Jesus himself would have owned no land. And so he sets out walking, a passerby, a man without property or prejudice, who saw the world as the kingdom of God.

Over the years, the literary critic Harold Bloom has offered some helpful insights into the Gospel of Thomas. In his afterword to Marvin Meyer's translation, Bloom writes this of Jesus' transitory ways: "As one who passes by, he urges his seekers to learn to be passersby, to cease hastening to the temporal death of the business and busyness that the world miscalls life. It is the busy world of death-in-life that constitutes the whatness from which we are being freed by the Jesus of the Gospel of Thomas." I like that phrase about the business and busyness that eat up so much of our everyday lives. It seems to me that is exactly what Thomas's Jesus is leading his followers away from. Beneath all of these distractions, we still might find the kingdom of God. And beneath all of the disguises we misidentify as the self, we still might find the light that at once transforms the self and the world into emanations of the Creator.

Because the Gospel of Thomas presents a portrait of Jesus so at odds with the canonical Gospels, if one wants to argue, as I do, for the primacy of *this* Christianity, then one must date Thomas closer its source, the talking Jesus, than any of the other four Gospels. There are many convincing reasons why the Gospel of Thomas is older than Mark's gospel. For starters, Thomas does not mention the destruction of the Second Temple. It seems unlikely that any Jewish writer would have ignored such a disastrous turn of events if he or she had lived through it, which suggests that Thomas was composed before 72 CE. Helmut Koester has also noted that Thomas must have come from an early group that was still appealing to the authority of the men who would become the Pillars of Jerusalem—Peter, James, and John. In Thomas 12, the "followers" say to Jesus, "We know that you are going to leave us. Who will be our leader?" Jesus says to them, "No matter where you are, you are to go to James the Just, for whose sake heaven and earth came into being."

Since James died in 62 CE, that would date Thomas's gospel well before Mark's. It is also worth noting here that in Thomas, Jesus has only followers, not the symbolic "twelve disciples" later revered by Matthew and Luke as symbolic counterparts to the twelve tribes of Israel. That, along with the fact that, in Thomas, Jesus is never called "Christ" or "Lord" or the "Son of God," suggests that those titles came later and were applied by the Pauline communities to whom the significance of Jesus' death had overshadowed and nearly replaced the message of his teachings. If Jesus had been the Messiah, if he had performed the miracles attributed to him by Luke, he certainly would have attracted some attention. He lived, after all, very close to large Roman cities like Tiberias and Sepphoris. But Jesus never did draw much attention. If he had, *some* Roman historian would have certainly taken notice. None did, which leads one to suspect that while Jesus may have had the shamanlike powers to exorcise psychic demons in people, he did not routinely walk on water and raise the dead. On the contrary, the Roman historian Tacitus wrote thus of the time when Jesus would have been wandering and teaching: "*Sub Tiberio quies*"—"Under Tiberius, nothing happened."

But the most substantial evidence for the early composition of Thomas lies in the sayings themselves as compared with the synoptic Gospels. When the Gospel of Thomas reemerged from Nag Hammadi, many scholars dismissed it as a late gospel that derived from the Gospels we all know. But Stephen Patterson, along with other scholars, has shown that Thomas resembles Matthew and Luke only when all three Gospels are independently using an earlier source—probably Q. When there are similar sayings in Thomas and the synoptic Gospels that don't rely on Q, the wording of the sayings differs too widely to suggest Thomas's dependence on the other Gospels. In addition, Thomas almost never orders his sayings in the same sequence as Matthew, Mark, and Luke, as he certainly would have done had they been his sources.

Perhaps the best argument for Thomas as the earliest account of Jesus' teachings is the way he simply lets them stand on their own, without adding an interpretation that would align Jesus with the theological biases of later writers. One might suppose that the compilers of Thomas and *Q knew* the context of the sayings, and saw no reason to replicate it, nor did they try to distort it to fit their own theological biases. For instance, Jesus' well-known remark about the impossibility of serving two masters stands alone in both Thomas and in Q. However, Luke, who would have borrowed it from *Q*, positions this saying within his narrative to function as a direct attack on the Pharisees. After Jesus says one cannot serve "God and mammon," Luke adds:

> The Pharisees, who were lovers of money, heard all this, and they scoffed at him. But he said to them, "You are those who justify yourselves before men, but God knows your hearts; for what is exalted among men is an abomination in the sight of God."
>
> (Luke 16:14–15, RSV)

The Pharisees were affiliated with the synagogues in the outlying towns. Mark directed his ire at the priests in Jerusalem. Because Mark understood the destruction of the Second Temple as a divine vendetta for the Crucifixion, notice the difference between Thomas 65 and Mark 12:1–11 in the parable of the vintner. Here is Thomas:

He said, "A . . . person owned a vineyard and rented it to some farmers, so that they might work it and he might collect its produce for them. He sent his servant so the farmers might give the servant the produce of the vineyard. They seized, beat, and almost killed his servant, and the servant returned and told his master. His master said, 'Perhaps he did not know them.' He sent another servant, and the farmers beat that one as well. Then the master sent his son and said, 'Perhaps they will show my son some respect.' Since the farmers knew that he was the heir to the vineyard, they seized him and killed him. Whoever has ears should hear."

(Thomas 65)

Mark tells the same story, but then adds his own ending:

What will the owner of the vineyard do? He will come and destroy the tenants, and give the vineyard to the others. Have you not read this scripture: "The very stone which the builders rejected has become the head of the corner; this was the Lord's doing, and it is marvelous in our eyes?"

(Mark 12:10–11, RSV)

In Mark's hands, the parable comes to mean that Jesus is the rejected son, and once the Temple is destroyed, he will return to build a new church—a Christian church. One sees a similar strategy at work in Matthew and Luke. Compare Thomas 8 with Matthew 13:47–50. Here is Thomas:

> Humankind is like a wise fisherman who cast his net into the sea and drew it up from the sea full of small fish. Among them the wise fisherman found a fine large fish. He threw all the little fish back into the sea and with difficulty chose the large fish. Whoever has ears to hear should hear.
>
> (Thomas 8)

Here is Matthew:

> Again, the kingdom of heaven is like a net which was thrown into the sea and gathered fish of every kind; when it was full, men drew it ashore and sat down and sorted the good into vessels but threw away the bad. So it will be at the close of the age. The angels will come out and separate the evil from the righteous, and throw them into the furnace of fire; there men will weep and gnash their teeth.
>
> (Matthew 13:47–50)

In the latter, not only do small and big fish become bad and good fish, but Matthew imposes his own interpretation on the parable, warning the reader to be a "good fish" that does not get cast into the fires of hell. It would not have occurred to the people of the Jesus movement to think in such terms. As the scholar Ron Cameron suggests, they would have understood it as a wisdom parable about "the discovery of one's own destiny."

Both passages from Mark and Matthew are about the death and the second coming of Jesus, two subjects that, as we have seen, are never raised in the Gospel of Thomas. As Helmut Koester wrote of Thomas and Q, "Both documents presuppose that Jesus' significance lay in his words, and in his words alone." Of course, the Gospel of John elevated the words of Jesus—his *logia*—to *the* Word, the *Logos*, which "was with God and was God." John's Word became flesh so as to mingle with humankind and ultimately save it from the wages of sin. Because John's gospel bears little resemblance to Matthew, Mark, or Luke, many scholars doubt its veracity as anything other than a philosophical treatise on the divine nature of Jesus.

John's Jesus is a divine savior, on his way to prepare a better place for those who believe in his redeeming power. Thomas's Jesus, as we have seen, is just the opposite, which would account for why early bishops would have ordered the

Gospel of Thomas destroyed. John preached that only Jesus carried the divine light; Thomas made explicit that everyone had the capacity to recognize the "light within." Elaine Pagels argues that because these two inheritors of Jesus' teaching had reached profoundly irreconcilable understandings of that message, particularly in regard to the kingdom of God, John actually wrote his gospel to refute Thomas's. Therefore only John's gospel depicts Thomas in a poor light—the one follower who doubted that Jesus had actually risen from the grave. In the end, of course, John's Jesus—the savior who could forgive sins and assuage our fear of death by promising an eternal afterlife—proved more attractive to Christians than Thomas's wandering mystic who called for voluntary poverty and spoke in maddening parables.

Unfortunately, along with John's "good news" comes the not-so-good news that we are, in a word, *guilty*—sinners by birth, consigned to serve out our sentence in this toilsome, fallen world. I can't say exactly when I lost my family's faith in a redeeming Messiah. Unlike the "conversion experience," I found that disbelief was not a cataclysmic event. But I do know that at some point during a lengthy bout of depression, accompanied by the acute self-loathing I learned in the Baptist church, I decided that I could best avoid my father's fate by abandoning my grandfather's beliefs. Of course losing faith is never that simple, and in my case it involved bitter recriminations, a lot of hurt feelings, and long, terrible silences. When my grandfather died a few years ago, we were barely speaking at all. And of course the problem with losing faith is that you never really do, not completely.

So when I first discovered the Gospel of Thomas, I was profoundly shocked and relieved to find a version of Christianity that I could actually accept—one that I felt might be a vital corrective to my family's view that we live helplessly, sinfully in a broken world. According to Thomas's Jesus, humankind never suffered an irredeemable Fall. The world only *appears* to be a realm of our separation from the Creator and from one another. When Thomas's Jesus tells his followers that "Adam came from great power and great wealth, but he was not worthy of you," he is implying that Adam's first sin was to take on the knowledge of good and evil—the knowledge that continues to divide the world into *us* and *them*. The stunning message of Thomas's gospel is that such divisions are arbitrary, destructive, and finally unnatural. Only the talking animals believe in them. Thus Adam's sin, ironically, was simply ignorance. True, that ignorance proved to be congenital, but it wasn't terminal, and it didn't demand divine intervention. We do not have to wait for the kingdom of God. We simply have to see it. Stephen Patterson put it nicely: "Thomas's Jesus is not really out to change the world. He aims rather to change people's perception of the world and their role in it."

This teacher of reconciliation was the same Jesus that Thomas Jefferson hoped to recover through his own gospel project. And whereas Jefferson found in Jesus' teaching an ethic for how we should treat others, Ralph Waldo Emerson found in it an alchemical light that transforms flesh into spirit. In some uncanny trick of history and geography, the ancient Gospel of Thomas combines these two visions of Jesus to give us what I would call a truly American Gospel. By pulling the kingdom of God out of the sky and transposing it onto this world, Thomas's Jesus returns us, in effect, to Jefferson's agrarian America where the farmer intuits the laws of God through the laws of nature. The Jesus of Thomas rejects John's apocalyptic End Time for an eternal present that observes only a natural calendar. Indeed, the Gospel of Thomas might be the first Christian text devoted to the subject of *ecology*. It calls us *back* to the natural world, back to a world that Yahweh first commanded Adam to "keep and cultivate." It is, alas, a gospel we Americans sorely need.

As the world all around us grows increasingly sicker, the Gospel of Thomas suggests that it is time we inverted Pascal's famous wager to say not that we should believe in heaven because we have nothing to lose, but rather that we should believe first in this world because in losing it we may lose everything. And if we can somehow live justly, modestly, with generosity and compassion, we have everything to gain. Perhaps we do not have to wait for the kingdom of God.

The End of Religion

"We tell ourselves stories in order to live." That is the first sentence of Joan Didion's landmark essay "The White Album." What the author intended to say, I think, is that stories carry meaning, meaning suggests order, and order implies purpose—a reason to believe, a reason to live. And perhaps we tell ourselves the very first story—the Creation story—to answer the two questions that rise into the collective consciousness of nearly every storytelling tribe:

Where did the world come from?

Why are we here?

The first question, we tell ourselves, might eventually be answered externally, if our telescopes are long enough, our microscopes powerful enough, and this is the domain of science. But to answer the second question, we must turn our instruments inward, toward the terrain of imagination, intuition, prophecy.

From what family friends tell me, my father struggled hard to answer this second question for himself. When my grandfather finished his fifty years in the ministry, the congregation at Bethel Baptist Church in Poquoson, Virginia, held a retirement party in the church's fellowship hall. My father had grown up just down the street really, close to my grandfather's first church, Riverside Baptist. At the reception, one of his friends, a man I had never met, handed me a glass of punch and started talking about my father. He clearly had something on his mind, something he wanted me to know. He had gone to high school with my father, and after graduation, they had both spent the summer contemplating the call of the ministry.

"Your father," he began, "he thought about things in ways no one else around here did. He really wrestled with these hard theological problems. I remember that summer, we would have coffee after supper at your grandparents' house, and we would spend hours discussing these things. Then your father would walk with me back to my house, and we'd make coffee again, and start the discussion all over. Then I would walk home with your father, and we'd make more coffee, and talk some more."

I could tell he thought this was information I needed, something that would confirm that my father was like me, or that I was like him. And I thanked the man for it. But ultimately, for all of my father's effort, the question *Why are we here?* is one he couldn't answer, at least not adequately enough to give *him* a reason to live. In the end, my father was locked inside my grandfather's story, the morality tale of fundamentalist Christianity. He couldn't find a way to make it his own, and a bullet to the head is the most emphatic form of punctuation—a statement, once and for all, that this sad story is over.

But these two questions raised by the Judeo-Christian creation story still have implications that go far beyond my personal story. Throughout this book, I have been circling around the first few chapters of Genesis. Now I want to consider the remarkable way it answers both questions. Chapter one takes up the scientific question of where the world came from and offers as an answer the epic week of creation. Thirty-five verses later, the story moves, in a much more sublime fashion, toward answering the second question of why we are here. Three thousand years later, astral physics and molecular biology have offered much more convincing answers to the first question. But as for question two, the Genesis account remains astonishingly relevant.

As the Bible scholar Richard Elliott Friedman has pointed out, when the people of ancient Judah began wrestling with that second question, the answer took the form of the world's first novel—a prose account of how this tribe came into being and eventually established a kingdom on earth. Unlike Homer's poetry, this work was not meant to be sung, and according to Friedman, unlike *The Iliad*, it was as likely as not written by a woman. Her task was complicated by the problem that the question actually took two forms within the minds of the Judean people: *Why are we here in the first place?* and then later *Why are we now here—east of Eden?* And it is clear from the story the Yahwist tells that the second question implied two more anxious queries: *Why do we suffer?* and *Why do we have to die?*

Here, then, are the significant details of her account of creation: The god YHWH (Yahweh) molded man (*adam*) from dust (*adamah*), breathed life into him, and built for him a garden paradise. Yahweh planted in the middle of this Garden of Eden two trees—the tree of life and the tree that yields the knowledge of good and evil. Then Yahweh issued his one prohibition:

"You are free to eat from any tree in the park," he said, "but you must not eat from the tree that yields knowledge of good and evil, for on the day you eat from that tree you shall die."

(Genesis 2:16–17, trans. James Moffatt)

Next God created woman from the man's rib. The stage was set. The principals took up their roles.

> Both of them, the man and his wife, were naked, but they felt no shame.
>
> (Genesis 2:25, trans. James Moffatt)

Then a serpent appeared, like the trickster in many other creation stories. He approached the woman, and inquired about restrictions on the property. Though she had only heard of it secondhand, the woman told him of the one proscription—the forbidden fruit.

> "No," said the serpent to the woman, "you shall not die; God knows that on the day you eat from it your eyes will be opened and you will be like gods, knowing good and evil."
>
> (Genesis 3:4–5, trans. James Moffatt)

What is striking about this passage—and what I have never heard a minister, certainly not my grandfather, acknowledge in front of a congregation—is that the serpent is right on both counts. He seems to have caught the Creator in a lie. Adam and Eve did not die when they ate from the forbidden tree, as their God Yahweh foretold, but instead lived to bear hardships east of Eden. The legend depends on this. Of course it is easy to argue that they did *eventually* die, and that this is what God meant. But still we are left with the literal translation of the Hebrew, which reads: "On *the day* you eat from that tree you shall die." Even more extraordinary than this divine slip is the realization that the serpent's epistemology is also accurate. Yahweh admits as much in a sublime confession. After Adam and Eve eat the fruit, He concedes to the angels:

> "Man has become like one of us, he knows good and evil. He might reach his hand now to the tree of life also, and by eating of it live for ever!"
>
> (Genesis 3:22, trans. James Moffatt)

We realize for the first time that the knowledge of death is derived from the knowledge of good and evil. Otherwise the first man and woman would certainly have eaten from the tree of life earlier; there was no prohibition against that. Of course they were immediately cast out of the Garden after they ate from the tree of knowledge, allowing the author to answer for her readers the question *Why must we die?* And she uses the same prop to justify human suffering. By eating from the forbidden tree, the first man and woman disobeyed Yahweh, and thus *deserve to suffer* as exiles from Paradise.

And in answering these two questions about life east of Eden, the author then returns to her fundamental question: *Why were we put here to begin with?*

It is through Yahweh's admission that the human beings have attained the divine knowledge of good and evil that the writer of this story unleashes its most subversive element, an element that, it seems to me, the Judeo-Christian tradition has always insisted on ignoring. Her answer to the fundamental question *Why are we here?* hinges on Yahweh's admission, "Man has become like one of us, he knows good and evil." From this, we can make only one inference as to why this world was brought into being: because Yahweh wanted to create a realm that *did not operate according to the laws of good and evil.* There is no other accounting for his single, arbitrary prohibition.

As I said earlier, when Adam and Eve ate from the forbidden tree, *values* were born. Language was no longer a poetry of naming as when Yahweh first marched the beasts past Adam and asked what he, *Homo loquax*, wanted to call them. Through such naming, Adam and Eve took part in the mystery of Creation, making words from the same substance— *pneuma*, breath—with which Yahweh called the world into being. But now that Adam and Eve were saddled with the new knowledge of good and evil, language no longer celebrated and affirmed the world, but instead judged it.

This is how I have come to understand the Judeo-Christian creation story. It is very different from the conclusions my grandfather reached in his sermon "Open Eyes." For him, the Primordial Parents had their eyes opened, not to the divine knowledge of good and evil, but to their own guilt. Whereas it seems to me that the serpent told Adam and Eve the truth, my grandfather believed the serpent lied "and has been lying to the human race ever since." In my grandfather's understanding, the Fall from God is so irrevocable that humankind can only be redeemed through a violent blood sacrifice. In his version, sin is no longer merely a "mistake," as the Hebrew translation has it; rather, sin is something so profoundly part of our nature that the flesh becomes synonymous with corruption. Only God himself, by taking the form of a sacrificial lamb, "his only son," can buy back his love for us.

My grandfather, of course, was not alone in believing this. Nearly every Baptist preacher of his and my generation believed and still believe it, along with most other Protestant and Catholic clergy. This is the fundamental theology that Europeans brought to North America, and it remains the core of mainstream Christianity. Unfortunately, because this Pauline Christianity places the kingdom of God in some unearthly hereafter, it too often ignores the damage that our behavior causes in this world.

As Americans, we have nothing in our collective behavior,

neither our foreign policy nor our consumer economy, that a Mediterranean street preacher named Jesus would recognize as part of his teaching. And if this militant status quo has become the American Religion, then it is in need of radical revision. We need to take the examples of Thomas Jefferson's Bible and the Gospel of Thomas, alongside the philosophies of William Byrd and Ralph Waldo Emerson, the poetry of Walt Whitman, and the science of Lynn Margulis, to imagine a new American Gospel that begins by returning the kingdom of God to this world, and us to it.

Whitman's remark that "whatever satisfies the soul is truth" strikes me as the first seed of America's only homegrown philosophy, pragmatism. And like the teachings of Jesus in the Gospel of Thomas and in Thomas Jefferson's gospel, pragmatism—particularly the writings of William James and John Dewey—is a philosophy that emphasizes the importance of what we do in *this* world, not the next. Pragmatism is not *opposed* to beliefs in the afterworld, but such beliefs are important only insofar as they spur action and bring about positive effects *here*. If we believe Jesus died for our sins but do not act in accordance with his teaching, then that belief ultimately amounts to very little. Conversely, if a belief in God prompts one to advocate for the dispossessed, then that belief should be trusted. James asked if it made a difference in the world whether one believed or did not believe something to be true, and he concluded that if one's belief in God has no repercussions in one's actions, or in the world, then it makes no difference whether one believes or not. "It is astonishing to see how many philosophical disputes collapse into insignificance," wrote James, "the moment you subject them to this single test of tracing a concrete consequence."

For James and Dewey, truth isn't found; it is *made*. Or as James put it, "Mind *engenders* truth *upon* reality." And truth is made by turning thought into action (in all of the Gospels, Jesus warned against trees that bear no fruit). Why contemplate the ultimately unknowable mysteries of metaphysics, asked the pragmatists, when there are serious problems in *this* world to contemplate and act upon? The pragmatist turns away from fixed principles, away from abstract definitions of truth and justice, back to the concrete world of facts and actions. As a case in point, Dewey's good friend Jane Addams put the philosophy of pragmatism into action in 1889 when she opened Hull House in Chicago's poor Nineteenth Ward. The settlement school provided art and educational programs, along with basic social services, to poor immigrants and minorities. Echoing Dewey and James, Addams wrote, "The ideal and developed settlement would attempt to test the value of human knowledge by action. . . .

The settlement stands for application as opposed to research; for emotion as opposed to abstraction."

Politically, pragmatism advocates two basic tenets: freedom of expression for the individual and justice for the community. The contemporary pragmatist Richard Rorty put it concisely: "[John Stuart] Mill's suggestion that governments devote themselves to optimizing the balance between leaving people's private lives alone and preventing suffering seems to me pretty much the last word." Like Whitman, John Dewey thought of the ideal American democracy as an "organism," a whole that is reflected in each of its parts—that is to say, in each individual citizen. The individual "embodies and realizes within himself the spirit and will of the whole organism." As cells work together within an organism, Dewey believed individuals worked together within a good society to achieve "a unity of *will*." In Dewey's view, as in Whitman's, "every man is a priest of God." The sovereign state is reflected in the sovereign individual. And because democracy is a social idea, it is also ethical by its very nature. Ethics, after all, begins with the realization that one is a social being and therefore part of a larger whole. A man living alone on an island requires no ethic. An aristocracy or a dictatorship could be run by only a few of its members. But the democratic organism needs all of its cells. And those cells must act simultaneously in their own interest and in the interest of the larger organism. Dewey declared that the terms in the French motto "Liberty, equality, fraternity" are not just words but "symbols of the highest ethical idea which humanity has yet reached." As Americans, we tend to emphasize our liberty at the expense of equality and fraternity. But one cannot have real democracy without the ethical components of equality, at least equal opportunity, and fraternal feelings of goodwill. Therefore, Dewey declared, "there is an individualism in democracy which there is not in aristocracy; but it is an ethical, not a numerical individualism; it is an individualism of freedom, of responsibility, of initiative to and for the ethical ideal, not an individualism of lawlessness." The individualism of freedom *and* responsibility constitutes Dewey's democratic ideal.

The genius of pragmatism is that it speaks at once to the experience of the individual and the experience of the community. If, as James said, the end—that is, the result—of religion should be "a richer, larger, more satisfying life," the end of pragmatism should be the same, both for the individual and the culture. I would contend that the individual realizes this goal through direct, authentic experience, and that the community, the larger culture, realizes it through collective, ethical actions.

To begin with the individual, Dewey emphasized the *quality* of experience. His study of aesthetics *Art as Experience* could just as easily have been called *Experience as Art*. "Experience," wrote Dewey, "is equivalent to art." That is to say, the highest form of experience should possess the same qualities as a work of art—aesthetic qualities. Dewey defined the aesthetic as "the clarified and intensified development of traits that belong to every normally complete experience." If, Dewey complained, the word "pure" had not been so denigrated by overuse, "we might say that aesthetic experience is pure experience." Within the pure experience, life takes on the qualities, the light, and the resonance of art when we step into the fullness, the immediacy, of the moment. And for Dewey, that moment is when experience takes on a religious quality. He made a clear distinction between the words "religious" and "religion." Religion is a set of prescriptions and prohibitions, and Dewey, raised in a family similar to mine, had no time for it. But the religious, he felt, was a quality, intuitive and resonant, that gives meaning to experience. In that sense, the religious and the aesthetic became as synonymous as art and experience. Each gave a sense of meaning to the other. Such deliberate forms of experience are the opposite of the restlessness that Tocqueville perceived in so many Americans. A passage in Henry David Thoreau's first book, *A Week on the Concord and Merrimack Rivers*, captures this sense of heightened experience. From their own boat, Thoreau and his brother John were watching two extremely skilled sailors guide their small skiff along the Concord River, and it caused Henry to muse:

> Their floating there was a beautiful and successful experiment in natural philosophy, and it served to ennoble in our eyes the art of navigation, for as birds fly and fishes swim, so these men sailed. It reminded us how much fairer and nobler all the actions of man might be, and that our life in its whole economy might be as beautiful as the fairest works of art or nature.

To raise one's own actions to a level where they would be worthy of contemplation—to turn life into art—was Thoreau's abiding wish, and it is the heart of Dewey's philosophy of the individual. Consequently, Dewey and James replaced Socrates' dictum *Know thyself* with the Emersonian charge *Invent thyself.*

In essence, what all of these writers are telling us is that art is a better interpreter of life than morality. Art, they are saying, is an emanation, a translation, of nature, whereas morality is an arbitrary, artificial code grafted onto our individual natures. To insert oneself into one's own story, as Whitman does in *Leaves of Grass*, is to avoid becoming trapped inside someone else's story of who you are, as happened to my father. For Thoreau and Dewey, men who had abandoned traditional Christianity, this meant pulling the kingdom of God out of the sky and imagining one's own life as a work of art. For me it meant leaving behind the Christian church—its sanctuary, its sermons, its deathly hymns—for what Thoreau called the "unroofed church" of the natural world. When I was in college, my grandfather convinced a fishing buddy to sell me an old canoe for cheap. I don't think he ever fathomed that it would take me so far from the church, the tradition, that gave so much meaning to his life. But the farther I paddled and drifted along the rivers of Virginia and Kentucky, the more I found myself gradually shedding the heaviness of my grandfather's story, and entering the milder estuaries of my own. My paddle took the measure of the moment—the measure of myself—and found nothing lacking. Sin would not drown me, guilt would not bind me to my family's tragic past. On those clear river days, I was baptized back into the kingdom my father and grandfather could never see, though it was spread out before them.

In *The Creation*, a book addressed to a Baptist minister much like my grandfather, E. O. Wilson makes the eerie prediction that future historians may remember our times as the Eremozoic era—the Age of Loneliness. While Wilson is referring mainly to our rapid loss of species, the metaphor can be extended to many aspects of our American lives. Not only are we estranged from the natural world, we are estranged from our most primal needs—the sources of our food, the sources of our clothes, and the sources of the energy that run our homes. Beyond these basic needs, money and corporate influence have estranged us from our own government. Our isolation within suburban homes has estranged us from our neighbors and our communities. Automated systems for exchanging money for goods and services have estranged us from the casual social contact that builds trust within communities. Cars have estranged us from our own bodies. And the often homogenous landscapes that cater to the automobile have estranged us from the local character of our native places. In his "Letter from Birmingham Jail," Martin Luther King, Jr., wrote that the definition of sin is "separation." He was speaking, of course, about the sin of racial segregation, but since then, many sins of separation and estrangement have lured us into the landscape of contemporary American culture—a landscape that has little to do with the Latin root *cultus*, meaning "cultivated" (as land) or "polished"; or the German term *Bildung*, referring often to cultivation of the self. The basic thrust of modernity has been to pull Creator, Creation, and humankind—what I've called the realized trinity—as far apart as possible. Now the problem—and the great potential—is that we can no longer keep this up. The average American needs 23.5 acres to produce those resources, when the sustainable level for everyone now alive on the planet is 4.5 acres. Our modern condition of estrangement has led us to live in ways that show a remarkable abdication of responsibility. And our estrangement from what Dewey called authentic experiences has led us too often into a way of conducting our days that is deadening to the soul.

Several different studies have recently shown that while the number of Americans who consider themselves happy stopped rising in 1975, real incomes have almost tripled since then. Our houses are twice the size they were in '75, and we consume twice as much as we did then. Still we seem no happier. In fact, the opposite seems true. We are more depressed, more anxious, more medicated, and seemingly more frazzled than at any other time in our history. William James diagnosed these problems years ago when he bemoaned "the moral flabbiness born of the exclusive worship of the bitch-goddess SUCCESS. That—with the squalid cash interpretation of the word 'success'—is our national disease." We have in many ways allowed the practice of accumulation to replace the practice of experience. The enemy of authentic, aesthetic experience, wrote Dewey, was what he called "the humdrum." More recently, the economist Tibor Scitovsky sought to figure out why Westerners with plenty of money still seemed so unhappy. His research turned up the same result as Dewey's— boredom, the humdrum. So if we cannot spend our way into happiness, and if the earth cannot absorb the waste and the cost of that spending, we might reconsider Jesus' advice in Thomas's Gospel: "Be passers-by." Such passing-by would bring back together those two pragmatic pillars—aesthetic experience (the personal) and ethics (the experience of the community). By passing by the realm of accumulation, we might pass back into more rewarding experiences. Almost everything that gave Thoreau pleasure—walking, writing, swimming, rowing (he built his own boat at age sixteen), reading, gardening, botanizing, socializing, singing—cost him nothing or next to nothing. Yet reading his journals, one senses that here was a man who lived an intensely rewarding life. Or consider a more recent example: There are now at least six thousand community gardens in the United States. And several decades of research have shown that in urban areas, PPI—people-plant interaction—results in reduced stress and anger, lower blood pressure and higher sense of self-worth for individuals, along with higher property values, reduced crime, increased business activity, and increased neighborliness. The beauty of this passage back into more rewarding, direct experience with the natural world is that it also carries us back into an ethic of responsibility, both to what the naturalist Aldo Leopold called the land community and to our own human communities. "The actual difference between individual*ism* and individuality of a true democracy," wrote Frank Lloyd Wright, "lies in the difference between selfishness and noble selfhood." The passing-by of Thomas's Jesus and the pragmatists is a directive away from the bored, acquisitive self, toward Wright's noble self. And because Wright, our greatest modern architect, offered a master plan for what a sustainable, ethical America should look like, a plan that is as ambitious as it is relevant, I will in a moment turn to his Jeffersonian treatise *The Living City*.

I met the writer Guy Davenport when I was a student at the University of Kentucky. Like me, Guy was a southerner who had been raised in the Baptist church. Gradually, mercifully, he became my mentor, and along the way we became friends. Guy used to call himself a "Baptist agnostic"—that is, he didn't believe in the church's promise of eternal salvation, but he was also under no illusion that he could wholly escape its influence. As a result, much of his fiction is a long working out of an alternative theology, one that looks very much like the sensual egalitarianism on display in Whitman's poetry.

I saw in Guy, I think for the first time in my life, someone who seemed completely alive. He believed unrepentantly in the beauty of the body and, equally so, the beauty of the well-crafted sentence. His house, bulging with books and paintings, was a kind of temple to art and literature. It was as inspiring a place to be as my grandfather's church was uninspiring. Guy was classically trained at Harvard and Oxford, but he most admired the American modernists, men and women like Ezra Pound, Gertrude Stein, and William Carlos Williams. His own work was fueled by that combination of tradition and originality. As a writer, Guy was wildly inventive, perhaps more than anyone of his generation. And he was *self*-inventive in the same way as Walt Whitman. He truly lived by Whitman's maxim that we are beholden only to the laws of our own nature. It didn't bother him that the neighbors thought him odd—he didn't own a TV, he didn't drive, and he could be seen daily walking around town with the kindling he had collected under his left arm ("free firewood") and a cigarette in his right hand. Once, outside a grocery store, a friend of mine mistook Guy for a street person; he had just won a MacArthur "genius" award. It might be going too far—and it might be unfair—to say that Guy, who was forty years older than me, came to represent an antidote to my grandfather's severe tutelage. But I did see in his example a way to find liberation from one's past in pursuit of the soul's idiosyncratic impulses.

Because Guy didn't drive, I took him Christmas shopping every year. We hit the malls and loaded up my truck; he was very generous toward his nieces and nephew, and his lifelong companion, Bonnie Jean. He appreciated the favor, I'm sure, but it didn't stop him from writing, in "Letter to the Masterbuilder," one of his last essays: "The automobile is a bionic roach. It eats cities." In our case, the city it had devoured was Lexington, Kentucky. Here, one-way streets ensure that traffic moves fast and pedestrians know their place. Downtown is, for the most part, a nondistinct collection of office buildings wrapped in reflective glass, where money changes hands during the day and homeless men and women wander the streets at night. Commerce, Guy surmised, has destroyed the American city as effectively as the British air force destroyed Dresden during World War II. "Conviviality (of the Mediterranean kind Le Corbusier said should be the *sole* purpose of a city) does not any longer exist," he wrote.

Guy's "Letter to the Masterbuilder" was modeled after a real letter that the nineteenth-century French utopist Charles Fourier sent to Napoleon. Western civilization, Fourier wrote, had screwed the pooch. First and foremost, it had failed to eliminate poverty. Moreover, four thousand years of Western philosophy had failed to bring about human happiness for one very important reason: no philosopher had ever simply accepted human beings as we are. After much reflection, Fourier wrote the emperor, he had solved both problems. His system of "passionate attraction" would accept rather than repress all natural passions—again, think Whitman—and it would guide men and women into work and associations best suited to their natural inclinations. So Fourier asked Napoleon for four hundred orphans (the French Revolution had produced plenty) with whom he would create a model for a new society—an "anti-civilization"—called Harmony.

Fourier's letter to Napoleon was intercepted by the police, who deemed its author "a harmless idiot," and Guy assumed (with some prescience, as we have recently found) that if he did write a letter to the president, "it would go directly to the FBI, with copies to the National Security Agency and the CIA." So instead he addressed his missive to a more generic figure, the Masterbuilder. Guy did not name his imperial architect, but I believe the mantle of the modern American Masterbuilder would have to go to Frank Lloyd Wright. Wright agreed (he couldn't have agreed more) with Guy that only the architect—more to the point, only Frank Lloyd Wright—could guide America into a happy future where democracy thrives and "mankind is free to function together in unity of spirit."

In 1958, the same year that James Robinson's group of scholars brought out the English translation of the Gospel of Thomas, Wright published the final version of his own utopian manifesto. It was issued first as *The Disappearing City* in 1932, when the Depression forced Wright to turn from building to writing, then as *The Living City* in 1958. I came across the latter some years ago in a used-book store, on a shelf labeled "Radical Thought." Now, on its fiftieth anniversary, *The*

Living City remains a not only radical but also extremely relevant vision of this country's future.

Like most utopian treatises, Wright's is a bit zany. His prose, filled with neologisms and grand pontifications on every subject, stands closest to the nutty, digressive prose style of Whitman's "Democratic Vistas." And indeed, Wright inherited from Whitman, Emerson, and William James the belief that human beings have hidden within us a divine nature, and that our greatest achievements come when the soul gets expressed in form or action. But whereas, at times, such soaring rhetoric can sound only like rhetoric in Emerson's essays, Wright *built* things. His ideas took concrete (literally concrete), brilliant forms. *The Living City* wasn't just a spiritual exercise Wright was putting himself through until new work came along. He meant for his plans to take shape, and through what he called his Broadacres project, Wright set out to recreate the American landscape, and in doing so, re-create and reinspire the American.

Wright charged, "Cities are huge mouths. Without the farmer, our towns and cities, big and small, would go naked and starve." He even predicted the crisis of peak oil: "Were motor oil and castor oil to dry up, the great big city would soon cease to function." And while the city was an unsustainable mammon, Wright believed the surrounding suburbs had robbed the citizen of his or her *individuality* and replaced it with a mass-produced, superficial *personality*. The American city and its suburbs had been corrupted by "sordid, ugly commercialism," he wrote, and as a consequence, selfish, "merchantile egotism" had replaced the democratic, Whitmanesque ideal of "noble selfhood." Style is character, Wright thought, and Americans had lost their style, traded it in, as Emerson charged a century earlier, for "conformity-culture." The once rugged American male had been emasculated by convenience, comfort, and commerce. This was partly because, within the industrial city, Americans had lost their connection to the natural world. Of our ancestral wanderer-forager, Wright determined, "On solid earth he was neither fool-proof nor weather-proof, but he was a whole man." Wright believed that capitalism was no more than a hangover of feudalism, where the poor still had to pay rent to the rich, who had often come by their wealth through speculation and lending, neither of which Wright approved of. Like the cranky Ezra Pound, Wright thought it a "monstrosity" that money could be used as its own commodity, rather than merely as an emblem of exchange. This was "plutocratic 'Capitalism'" at its worst. It bred inequality, poverty, malaise, and selfishness. And it had to go.

The solution, Wright felt sure, was to strip away the scaffolding of Alexander Hamilton's centralized system of banks, wealth, and power, and return the citizen, whom Wright often called the "usonian," to his or her agrarian, Jeffersonian roots. In a 1785 letter to Bishop James Madison, Thomas Jefferson wrote: "It is not too soon to provide by every possible means that as few as possible shall be without a little portion of land." Following suit, Wright argued that poverty could best be cured by giving every American at least one acre as a birthright. On his acre, this new homesteader, this urban agrarian, could slowly build a house out of sectional units (designed by Wright, of course) as he could afford them, and plant "shade trees, fruit trees, berry bushes, vegetables and flowers in the gardens." Instead of living in urban, public housing "about as inspiring as a coffin," poor families would find natural wealth and inspiration on "land they were born to inherit as they were born to inherit air to breathe, daylight to see by, water to drink."

Of course, if an individual or a family did not *want* to homestead, there were plenty of apartments in Wright's model, where usonians would still be close to ponds and parks. The point was to bring the terms "nature" and "culture" back into conversation. Wright agreed with Fourier and Whitman that happiness was the best gauge of a culture. Wright saw his fellow American conducting "a vicarious life virtually sterilized by machinery" and concluded that he or she was not happy. Wright's plan, harking back to Jefferson, was to "reawaken instincts of the agrarian," and in doing so, return character and style to both the individual and the city. "Our capitalist morals are not on speaking terms with Nature," he observed. Before E. O. Wilson developed his Biophilia Hypothesis, which states that human beings have an innate affinity for all of life, Wright proposed a return to nature that also brought the best parts of the city with it. The 1958 edition of *The Living City* comes with a large foldout map of Broadacres that shows cinemas and "music gardens" alongside the lakes, vineyards, and orchards.*

Like Emerson and Whitman before him, Wright spoke with a heraldic urgency to the larger themes of soul and civilization. And in a civilization shaped by commerce and industrialism, he believed that nothing less than our souls were at stake. But there was still a way out, because, he wrote, "the kingdom of God is within *you.*" This creative impulse Wright called the Poetic Principle, and it could, if we let it, guide us into a new future: "Our civilization might well be eternal instead of, as now, on the way to join the great rubbish heap of

---

*Not all of Wright's prescriptions proved tenable. His power of prophecy particularly failed him when he predicted that the automobile could be "humanized" into an unobtrusive kind of pod, a vehicular second skin. These streamlined cars would carry usonians to highway crossings where local merchants sold their wares at roadside markets. Wright didn't foresee that his sleek "body car" would become the stretch Hummer, or that interstate exits would instead be dominated by fast-food chains and Hustler stores.

the civilizations which history catalogues, if this true man-light—the Poetic Principle—would dawn afresh for us in such organic character." I believe this Poetic Principle has run through all of the American geniuses I have evoked in this book. What's more, each of them has expressed it as an *organic principle* because it is striving toward wholeness—a unity made strong by multiplicity. For Emerson and James, this meant an organic philosophy, and for Whitman it meant an organic poetry; Dewey called for a truly democratic, organic politics wherein each member is vital to the whole; Wes Jackson urged organic farming as an antidote to industrial agriculture, and Wright talked endlessly about a form of organic architecture "that builds from the inside out."

I have tried to make the case that all of these poets, scientists, philosophers, and builders do provide such inspiration to overcome the status quo, the corporations, the fatalistic fundamentalism of many churches, the poorly planned cities and suburbs, and the imperial direction of our federal government. In short, they offer the inspiration and the direction to avoid Wright's rubbish heap of civilizations. In his influential book *Collapse*, Jared Diamond offers a preponderance of evidence that overconsumption and overpopulation are the surest predictors of a civilization's impending collapse. In the fifth century, the Mayans were so busy building monuments to themselves that they couldn't see, until it was too late, that they had ruined their forests and rivers. "Why," Diamond asks, "did the kings and nobles not recognize and solve these problems? A major reason was that their attention was evidently focused on the short-term concerns of enriching themselves, waging wars, erecting monuments, competing with one another, and extracting enough food from the peasants to support all those activities." That we, the latest conquerors of North America, are doing the same thing was not lost on Frank Lloyd Wright. Though he knew nothing of climate change, he saw clearly that an America of centralized wealth and power was steering the country toward the same fate as the Mayans'.

In *The Living City*, Wright set forth, in philosophical and architectural details, a plan that would steer us away from collapse. He summed it up thus: *Decentralize and reintegrate.* Wright believed that if politics was decentralized, Americans would again take up their roles as active citizens, real democrats. If cities were decentralized, an organic architecture, "a natural interpreter of Nature," would rise to reflect the true character of particular places ("When man builds 'natural' buildings naturally, he builds his very life into them," Wright observed). If agriculture was decentralized, food would be fresher and farmers could deal directly, and more profitably, with consumers (and we would rely far less on fossil fuels). If the economy was decentralized, real need, as opposed to advertising, would drive production, and quality would replace quantity as the new standard of wealth. Everywhere, this return to the local would mean the end of urban and suburban anonymity and isolation, and the reintegration of people within distinct, independent communities. The result would be a confederation that Wright called "the integrated society of small units each of the highest quality imaginable."

Beginning with the writings of William Byrd and Thomas Jefferson, and leading up to the biological theories of Lynn Margulis, we find an American Gospel that equates nature with virtue, that finds in the land ethic a model for our human communities. Jefferson thought the farmer was the most virtuous of men simply because he worked closest to the Creation, and therefore, closest to the laws of the Creator. Under her microscope, Margulis observed cooperation on the most basic level of life—that of the cell. And E. O. Wilson, through his Biophilia Hypothesis, sees a psychology behind Margulis's biological findings by showing that we all have innate (if technologically obscured) affections for the natural world.

It was only very late in the game of accelerated evolution, which we call civilization, that psychology separated itself from religion and that religion separated itself from nature. It is my contention that through the realized trinity of Creator, Creation, and humankind, they can be reconciled; we as a culture can begin to replace the unsustainable, linear, industrial economy with an economy of nature. In trying to imagine the reconciliation, the reattachment of human beings to the Creation and the Creator, we must move beyond Aristotle's solely humanistic ethics, and think instead about what Aldo Leopold called a *land ethic.* "Examine each question in terms of what is ethically and esthetically right, as well as what is economically expedient," wrote Leopold. "A thing is right when it tends to preserve the integrity, stability and beauty of the biotic community. It is wrong when it tends otherwise." The model for such an integrated, stable, and beautiful system of communities is, I believe, the forest and its watershed. From the level of the canopy, down through the understory, to the shrub layer, the herb and fern layer, and finally the soil layer of decomposing leaves, forests form intensely symbiotic communities of flora and fauna. Bees pollinate wildflowers and fruit trees; leaf shredders serve up food at the headwaters of streams; termites break down dead logs; worms aerate soil around plant roots; squirrels distribute acorns. Each species finds its niche and learns to adapt to its surroundings. Each species depends on another and so has a stake in maintaining the health of the entire forest community, the entire ecosystem. In the forests closest to where I live, the mixed mesophytic, seventy different species of trees have

learned to coexist interdependently, sharing light and resources. Hemlocks and beech keep to the streams, oak and
hickories share the middle elevations, and pines form the
high, dry crowns of the ridgetops. In addition, watersheds are
models of Emersonian self-reliance. By their geographical
nature, watersheds must be self-sufficient. No other watershed is going to loan out water or fertile soil. So energy is
constantly being transferred from sun to plants to animals to
soil. All waste is food for next year's saplings.

What's more, forests do all of the work that modern cities
cannot: they sequester carbon, produce oxygen, purify air
and water, prevent flooding, and hold soil in place. This
should give us serious pause when comparing industrial
urban design with the more healthy and efficient natural laws
of the forest. The progressive architects and designers William McDonough and Michael Braungart have adopted the
latter in their own work. When they worked with the writer
David Orr at Oberlin College to build the Lewis Center for
Environmental Studies, they conceived of the building func-
tioning "the way a tree works." This meant a building that
could purify air, create shade, enrich soil, change with the
seasons, and create all of the energy it needed to operate. So
the design team put up solar panels (man's best attempt at
photosynthesis), built classrooms facing the south and west
for maximum exposure to sunlight, dug a pond to store water

for irrigation, and installed a "living machine" where a series of plant-filled ponds naturally purify wastewater. "Imagine a building like a tree, a city like a forest," McDonough and Braungart write in their book *Cradle to Cradle*. It's a sentence that could have easily been lifted from Wright's *The Living City*. What does such an economy look like? It maintains and even adds to the health of its watershed. It values diversity over homogeneity, in agriculture, physical landscapes, and human communities. It uses renewable natural sources of energy, not sources that have been stored in the ground for millions of years. It is interdependent and therefore, by definition, at once organic and democratic. It retrieves human experience from the passive, vicarious realm of similitude, the virtual, and returns it to the actual world.

When challenged by the Herodians as to whether the people of Jerusalem should pay taxes to Caesar, Jesus famously asked to see a Roman coin. When the questioner produced one that bore the profile of Augustus, Jesus remarked, "Render unto Caesar the things that are Caesar's, and unto God the things that are God's." Now that coin has become the American dollar, and that dollar has financed a corrupt political system at home, a Roman-style military presence around the world, and an industrial economy that we can no longer sustain. It is finally time, then, that we abandon the economy of Caesar and turn to the natural economy that, in this anecdote, Jesus offered as its alternative. In the early 1790s, George Washington's two most brilliant protégés, Thomas Jefferson and Alexander Hamilton, presented the president with two profoundly different visions for the country. Washington wavered, but ultimately chose Hamilton's urban vision of banks, national debt, and a strong federal government over Jefferson's agrarian vision of self-sufficient, self-governing wards. Washington thought, perhaps correctly, that Hamilton's was the more practical plan. But even if one entertains the premise that Washington was right to steer the nation by Hamilton's star, two centuries later, the grim reality is that we simply cannot remain on our myopic, unsustainable course.

In 1905, Henry Adams made the prescient observation that human civilization had begun careening into the future at an increasingly accelerated rate, following what he called the law of acceleration. Adams further theorized that the rate of acceleration could be measured by the amount of coal burned to drive this industrial juggernaut. The formula proved extremely accurate. Today, we burn seven million more tons of coal than just forty years ago. Combine the law of acceleration with the particularly American law of accumulation, and the result is alarming—so alarming it prompted E. O. Wilson to write an essay called "Is Humanity Suicidal?" It is here, on the subject of suicide, that my dual narrative—the story of my own family and my reading of American history—comes to a close.

As I wrote toward the beginning of this book, I believe my father took his own life in large part because he could not find *accommodation* within this world. I believe he took too seriously the old Baptist hymn "This World Is Not My Home." That message, and hundreds of similar sentiments preached from my grandfather's pulpit, caused my father to develop what William James called the sick soul, and that sickness ultimately drove him to suicide. The problem with that message, I have slowly come to realize, is not only that it can cause irreparable damage to the individual, but that it too often allows us to abdicate our responsibility to the Creation as a whole. It does not, like Walt Whitman's poetry, celebrate the natural world and, by extension, our own physical natures. Rather, it reduces the flesh to sin, and the natural world to, as Anne Bradstreet wrote in her 1678 poem "The Flesh and the Spirit," the "trash which earth doth hold." Like the Spirit in Bradstreet's poem, who hates "sinful pleasure" and whose "ambition lies above," we have become doubly estranged from this world of pleasure and belonging because of the abnegating theology that my grandfather and so many other American religious leaders inherited from the Puritans.

My father's story ended badly. It remains to be seen if our culture and our species is accelerating irreversibly toward suicide. We are not, of course, suicidal in any literal sense. But, says Wilson, we often *act* as if we are through a shortsighted way of living that once served, but no longer serves, our evolutionary needs. I believe the solution, the way to amend the law of acceleration, is to find what my father could not: accommodation. The word's Latin root, *accommodatus*, means "fit" or "suitable," and ultimately derives from *modus*, "measure." And *measure*—a deliberate, thoughtful accounting of our actions—is precisely what we must now find. We must, in absolute seriousness, adopt Frank Lloyd Wright's charge to *decentralize and reintegrate*. We must decentralize the production and transportation of our food, the distribution of energy, and the practice of politics. All supply lines that are powered by fossil fuels must be drastically shortened. We must begin to replace—not in one sweeping move, but in a thousand different ways—Hamilton's industrial economy for Jefferson's economy of nature. In the process, we will overcome our estrangement from the most basic elements of our lives and find ourselves reintegrated within communities that reflect the symbiosis, the interdependence, and the health of a forest watershed.

I have not quite come around to Tolstoy's belief that one must choose which kingdom to believe in—the heavenly kingdom or the kingdom that is right before us. I still cast my lot with the pragmatists and hold that if a belief in a heavenly kingdom causes one to act on the *teachings* of Jesus in this world, then one has no right to question such a belief. But I have come to understand that the kingdom of the Creation—*this* world—is where we must first discover evidence of the Creator, the presence of the holy, and I have come to see that the Creation is where we must first enact the teachings that Thomas Jefferson set before us in his own cut-and-paste gospel. It would seem an uncomfortable reality for many American Christians that, throughout the Gospels of the Bible, Jesus spent so much time talking about the poor and so little time (none, actually) talking about homosexuality, prayer in school, or where to post the Ten Commandments. Yet his concern for the poor remains distressingly relevant in light of recent meteorological research that shows the climate crisis will effect the world's poorest the most. What's more, it is the abuse of resources in the richest, most avowedly Christian nations that has caused and will continue to cause droughts, fires, famines, floods, and hurricanes in the parts of the world where the people played virtually no role in causing our global warming. Rajendra Pachauri, the chairman of the UN's Intergovernmental Panel on Climate Change, summarized the situation: "It's the poorest of the poor in the world, and this includes poor people even in prosperous societies, who are going to get the worst hit." There is no way we can demand that developing nations curb their use of fossil fuels if we are unwilling to change and do what we Americans seem so stubbornly resistant to—taking responsibility for our actions.

We Americans do not tend to change our behavior simply because we are told that we should. No real political, cultural, or environmental change—the crucial paradigm shift this country needs—will come about without a fundamental change of heart, an inspired, poetic desire to begin inventing the American future that Jefferson, Whitman, and Wright proposed. In the Gospel of Thomas, Jesus tells the wealthy woman Salome, "If one is whole, one will be filled with light," and he expresses a very pragmatic view of religion, that it should be a *habit of action*, when he says, "For no one lights a lamp and puts it under a basket, nor does one put it in a hidden place." Emerson called this inner light Intuition, as opposed to the tuitions whereby others tell us what to do. Whitman called it the soul. Frank Lloyd Wright called it the "man-light." Under whatever name, I believe it is the same impulse; it comes from the same mysterious but unmistakable source. This internal impulse must also be a force for change. These thinkers point to what is right in front of us, and then show how we might enter into that pure experience, that wholeness of being—the kingdom of God. They return us to what Whitman called "reason's early paradise," where we were installed by the Creator to care for the natural world, and to affirm, through language and action, its beauty, its stability, its integrity.

It has been my contention throughout this book that Jefferson, Ralph Waldo Emerson, Walt Whitman, and the pragmatists were far truer to the spirit of the teachings of a Mediterranean street preacher named Yeshua, or Jesus, than were the Puritans and the American religion that they engendered in this country. These thinkers, in their remarkable and various ways, all stayed true to the earliest Gospel we have, the one that they had no way of knowing, the Gospel of Thomas. They all intuited what Jesus said in the prologue to Thomas: "The kingdom is inside you and it is outside you." To discover one is to discover the other. What all of the authors of the American Gospel tell us is the same thing Jesus told his followers in the Gospel of Thomas: *There never was a Fall,* and therefore, *we do not need to be saved by a sacrificial martyr.* We only see it that way. We tricked ourselves into believing in the Fall so that suffering and death would make sense, so we could still believe in a just God who punishes us only because we disobeyed him and therefore deserve it. But the fact of the matter is that the author of Genesis never mentions a Fall. The truly radical message of the Genesis creation story is that we were installed here precisely to live *outside the laws of good and evil.* We were intended, rather, to follow what Whitman called the Laws for Creation. In a poem by that title, Whitman exhorted his fellow Americans:

> What do you suppose creation is?
> What do you suppose will satisfy the soul, except to
>   walk free and own no superiors?
> What do you suppose I would intimate to you in a
>   hundred ways, but that man or woman is as good
>   as God?
> And that there is no God any more divine than
>   Yourself?
> And that that is what the oldest and newest myths
>   finally mean?
> And that you or any one must approach creations
>   through such laws?

The oldest and newest myth is that the natural world is a holy realm where "there is no God any more divine than Yourself" because there is no separation among Creator, Creation, and Yourself. In the wake of this stunning reconciliation— the kind that Emerson said only the great poet can pull off— John Dewey and William James proclaimed that the work of philosophy should no longer concern itself with absolute definitions of good and evil, but rather should get back to the work *of* this world—namely, widening sympathy and reducing cruelty, and acting as if this world *is* in fact our home.

As an alternative to the mainstream American religion that admonishes us to be suspicious of our own nature, the authors of the American Gospel extol us to be true to our own nature, and to find that kingdom within writ large in the kingdom that is before us—the natural world. The greatest recommendation for a sustainable paradigm shift in this country, and the greatest chance that it will succeed, is the realization that authentic experience—direct experience of the body, the land, and the community—is more rewarding, more soulful than all of the poor substitutes offered up by our popular and passive consumer culture. To adopt the teachings of Thomas Jefferson's gospel and the Gospel of Thomas would mean to reinvent oneself and to reinvent the American landscape, both rural and urban, in ways that are truly more meaningful, more reverent, more just. We might discover that the natural world and our own natures are two mirrors that infinitely reflect the kingdom of God. There we might find both our refuge and our calling.

Sources

Introduction

Tolstoy, Leo. *The Kingdom of God Is Within You.* Trans. Constance
Garnett. Lincoln: University of Nebraska Press, 1984.
Troyat, Henri. *Tolstoy.* Garden City, NY: Doubleday, 1967.

Open Eyes

Dabney, Virginius. *Virginia: The New Dominion.* Garden City, NY:
Doubleday, 1971.
Price, David A. *Love and Hate in Jamestown: John Smith, Pocahontas,
and the Start of a New Nation.* New York: Random House, 2005.

William Byrd's New World

Beatty, Richmond Croom. *William Byrd of Westover.* Hamden, CT:
Alchon Books, 1970.

Bercovitch, Sacvan. *The Puritan Origins of the American Self.* New Haven, CT: Yale University Press, 1975.

Byrd, William. *The Great American Gentleman: The Secret Diary of William Byrd of Westover 1709–1712.* New York: Capricorn Books, 1963.

———. *The Prose Works of William Byrd of Westover.* Ed. Louis B. Wright. Cambridge, MA: The Belknap Press of Harvard University Press, 1966.

———. *A Secret History of the Dividing Line Betwixt Virginia and North Carolina.* Ed. William K. Boyd. New York: Dover, 1967.

Diamond, Jared. *Collapse.* New York: Viking, 2005.

James, William. *The Varieties of Religious Experience.* New York: Collier Books, 1961.

Nash, Roderick Frazier. *Wilderness & the American Mind.* New Haven, CT: Yale University Press, 1967.

Silverman, Kenneth. *The Life and Times of Cotton Mather.* New York: Harper & Row, 1984.

Williams, William Carlos. *In the American Grain.* New York: New Directions, 1925.

The Gospel According to Thomas Jefferson

Chernow, Ron. *Alexander Hamilton.* New York: The Penguin Press, 2004.

Ellis, Joseph J. *American Sphinx: The Character of Thomas Jefferson.* New York: Alfred A. Knopf, 1996.

Hogeland, William. "Inventing Alexander Hamilton." *Boston Review.* November/December 2007.

———. *The Whiskey Rebellion.* New York: Scribner, 2006.

Jefferson, Thomas. *The Jefferson Bible.* Ed. F. Forrester Church. Boston: Beacon Press, 1989.

————. *The Life and Selected Writings of Thomas Jefferson.* Eds. Adrienne Koch and William Peden. New York: Random House, 1944.

————. *Notes on the State of Virginia.* Ed. William Peden. New York: W. W. Norton, 1972.

Malone, Dumas. *Jefferson and the Rights of Man.* Boston: Little, Brown, 1951.

Wilson, E. O. *The Creation.* New York: W. W. Norton, 2006.

Walt Whitman at Furnace Mountain

Dae Gak. *Going Beyond Buddha.* Boston: Charles E. Tuttle, 1997.

Dawkins, Richard. *River Out of Eden: A Darwinian View of Life.* New York: Basic Books, 1995.

Emerson, Ralph Waldo. *The Complete Essays and Other Writings.* New York: Random House, 1940.

James, William. *Writings 1878–1899.* New York: Library of America, 1992.

Jung, Carl. "The Difference Between Eastern and Western Thinking." In *The Portable Jung.* Ed. Joseph Campbell. New York: Viking, 1971.

Margulis, Lynn. *Microcosmos.* Berkeley: University of California Press, 1997.

————. *Symbiotic Planet.* Boston: Basic Books, 1998.

Merton, Thomas. "Symbolism: Communication or Communion?" In *New Directions 20.* Ed. J. Laughlin. New York: New Directions, 1968.

Rexroth, Kenneth. *One Hundred Poems from the Chinese.* New York: New Directions, 1971.

Reynolds, David S. *Walt Whitman's America: A Cultural Biography.* New York: Random House, 1995.

Richardson, Robert D., Jr. *Emerson: Mind on Fire.* Berkeley: University of California Press, 1995.

Suzuki, D. T. *Zen Buddhism.* New York: Anchor Books, 1956.

Whitman, Walt. *Complete Poetry and Selected Prose by Walt Whitman.* Ed. James E. Miller, Jr. Boston: Houghton Mifflin, 1959.

———. *Leaves of Grass.* Ed. Malcolm Cowley. New York: Viking Penguin, 1959.

The Kingdom of God

Borg, Marcus J. *The God We Never Knew: Beyond Dogmatic Religion to a More Authentic Contemporary Faith.* San Francisco: HarperSanFrancisco, 1997.

Brown, Lester R. *Plan B 3.0: Mobilizing to Save Civilization.* New York: W. W. Norton, 2008.

Crossan, John Dominic. *The Historical Jesus: The Life of a Mediterranean Jewish Peasant.* New York: HarperCollins, 1991.

———. *Jesus: A Revolutionary Biography.* New York: HarperCollins, 1994.

Davies, Stevan, trans. *The Gospel of Thomas.* Boston: Shambhala, 2004.

Kloppenborg, John S., Marvin W. Meyer, Stephen J. Patterson, and Michael G. Steinhauser, eds. *Q Thomas Reader.* Sonoma, CA: Polebridge Press, 1990.

Koester, Helmut. *Ancient Christian Gospels.* Philadelphia: Trinity Press International, 1990.

Mack, Burton L. *The Lost Gospel.* San Francisco: HarperSanFrancisco, 1993.

———. *The Myth of Innocence.* Philadelphia: Fortress Press, 1988.

———. *Who Wrote the New Testament?* San Francisco: HarperSanFrancisco, 1995.

Meyer, Marvin, trans. *The Gospel of Thomas: The Hidden Sayings of Jesus.* San Francisco: HarperSanFrancisco, 1992.

Pagels, Elaine. *Adam, Eve, and the Serpent: Sex and Politics in Early Christianity.* New York: Random House, 1988.

———. *Beyond Belief.* New York: Random House, 2005.

———. *The Gnostic Gospels.* New York: Random House, 1979.

Patterson, Stephen J. *The God of Jesus.* Harrisburg, PA: Trinity Press International, 1998.

———. *The Gospel of Thomas and Jesus.* Sonoma, CA: Polebridge Press, 1993.

Patterson, Stephen J., James M. Robinson, and Hans-Gebhard Bethge, eds. *The Fifth Gospel: The Gospel of Thomas Comes of Age.* Harrisburg, PA: Trinity Press International, 1998.

Robinson, James. *The Gospel of Jesus.* San Francisco: HarperSanFrancisco, 2005.

The End of Religion

Addams, Jane. "A Function of the Settlement School." In *Pragmatism: A Reader.* Ed. Louis Menand. New York: Vintage, 1997.

Alperovitz, Gar. "California Split." *The New York Times,* February 10, 2007.

Bloom, Harold. *The Book of J.* New York: Grove Weidenfeld, 1990.

Bourne, Randolph S. *War and the Intellectuals.* New York: Harper & Row, 1964.

Davenport, Guy. "A Letter to the Masterbuilder." In *The Hunter Gracchus.* Washington, DC: Counterpoint, 1996.

———. "The Symbol of the Archaic." In *The Geography of the Imagination: Forty Essays by Guy Davenport.* San Francisco: North Point Press, 1981.

Dewey. John. *Art as Experience.* New York: Perigee Books, 1980.

———. *A Common Faith.* New Haven, CT: Yale University Press, 1934.

———. "The Ethics of Democracy." *Pragmatism: A Reader.* Ed. Louis Menand. New York: Vintage, 1997.

———. "The Need for a Recovery of Philosophy." *Pragmatism: A Reader.* Ed. Louis Menand. New York: Vintage, 1997.

Diamond, Jared. "The Last Americans." *Harper's,* June 2003.

James, William. *Pragmatism.* Buffalo, NY: Prometheus Books, 1961.

Jung, Carl. "The Difference Between Eastern and Western Thinking." *The Portable Jung.* Ed. Joseph Campbell. New York: Penguin, 1976.

Layard, Richard. *Happiness: Lessons from a New Science.* New York: The Penguin Press, 2005.

Leopold, Aldo. *A Sand County Almanac.* New York: Ballantine, 1970.

Mack, Burton L. *A Myth of Innocence: Mark and Christian Origins.* Minneapolis: Augsburg Fortress, 1998.

Malakoff, David. "What Good Is Community Greening?" American Community Gardening Association. www.communitygarden.org/docs/learn/articles/WhatGoodisCommunityGreening.pdf.

Manning, Richard. "The Oil We Eat," *Harper's,* February 2004. (Source for quotations of David Pimentel.)

McDonough, William, and Michael Braungart. *Cradle to Cradle.* New York: North Point Press, 2002.

Thoreau, Henry David. *A Week on the Concord and Merrimack Rivers.* New York: Penguin, 1998.

Wilson, Edward O. *Biophilia.* Cambridge, MA: Harvard University Press, 1984.

———. *The Creation: An Appeal to Save Life on Earth.* New York: W. W. Norton, 2006.

Wirzba, Norman. *The Paradise of God.* Oxford: Oxford University Press, 2003.

Worster, Daniel. *Nature's Economy.* San Francisco: Sierra Club, 1977.

Wright, Frank Lloyd. *The Living City.* New York: Horizon Press, 1958.